"This book draws on a dazzling array of conceptual resources to explore the ethical issues of integrating future artificial agents into communities in ways that may challenge the very meaning of human flourishing."

Matthew Dennis, *Eindhoven University of Technology, The Netherlands*

W0091464

Hybrid Societies

This book explores how social robots and synthetic social agents will change our social systems and intersubjective relationships.

It is obvious that technology influences societies. But how, and under what conditions do these changes occur? This book provides a theoretical foundation for the social implications of artificial intelligence (AI) and robotics. It starts from philosophy of technology, with a focus on social robotics, to systematically explore the concept of socio-technical change. It addresses two main questions: To what extent will social robots modify our social systems? And how will human relationality be affected by human–robot interactions? The book employs resources from continental philosophy, actor–network theory, psychoanalysis, systemic theory, and constructivist cognitive theory to develop a theory of socio-technical change. It also offers a novel perspective on how we should evaluate the effectiveness of social robots, which has significant implications for how social robotics should be researched and designed.

Hybrid Societies will appeal to scholars and advanced students working in philosophy of technology, AI ethics, robot ethics, and continental philosophy.

Piercosma Bisconti is researcher at the Italian Interuniversity Consortium for Computer Science and is founder of the company DEXAI – Artificial Ethics, and representing Italy in the standardisation processes of AI in ISO and CEN–CENELEC. His research engages with the social and ethical implications of artificial intelligence and robotics.

Routledge Studies in Contemporary Philosophy

For more information about this series, please visit: www.routledge.com/Routledge-Studies-in-Contemporary-Philosophy/book-series/SE0720

Hybrid Societies
Living with Social Robots

Piercosma Bisconti

Routledge
Taylor & Francis Group

NEW YORK AND LONDON

First published 2024
by Routledge
605 Third Avenue, New York, NY 10158

and by Routledge
4 Park Square, Milton Park, Abingdon, Oxon, OX14 4RN

Routledge is an imprint of the Taylor & Francis Group, an informa business

ISBN: 978-1-032-60592-0 (hbk)
ISBN: 978-1-032-60590-6 (pbk)
ISBN: 978-1-003-45979-8 (ebk)

DOI: 10.4324/9781003459798

Typeset in Sabon
by Newgen Publishing UK

Contents

Acknowledgements

To remain faithful to the content of this manuscript, I would like to begin by thanking my desk. Not all desks inspire the same thoughts or bear the same ideas. The solidity of this one's plan and the balance of its supports have, I hope, both flowed into this manuscript.

Among the many people I would like to thank for supporting me during these years, some must receive special mention.

To Nicoletta, I owe the systemic theoretical backbone driving this work.

Any idea contained in this book has been discussed countless times and from every possible point of view with Federico, lifelong friend, to the point that I cannot tell to which of us this essay's arguments belong.

Valeria C. helped shape the fundamental ideas on socio-symbolic systems, along with my past and present passion for political philosophy and my firm conviction that there is no theory without practice.

Since childhood, my father has given me an immeasurable zeal for conceptual investigation, always reminding me that the more slippery the theoretical ground, the greater the rigour required. I hope I have learned this lesson.

To Prof. Barbara Henry I owe inexhaustible gratitude for having believed in my research project when no one else seemed to do so. I hope I have kept faith with this trust, without which this journey would never have started.

To Valeria M. I owe gratitude for pushing me to follow what I wanted to do when it no longer seemed possible.

Finally, I would like to thank cigarettes. There is no social mediator more effective than smoking for smoothing social interactions between human beings.

1   Introduction to social robotics

Since the dawn of history, humans have been fascinated by the αὐτόματος, a Greek word meaning "a thing that moves by its own will". An αὐτόματος is an object, not a living being, that appears – unbelievably – not to be tied to the strict laws of causality, instead behaving autonomously within its environment. Philosophy first, and science thereafter, studied the possibility of transmuting cold matter into an apparent life form, and the creation of an αὐτόματος has always represented the final achievement of this endeavour. In the last hundred years, technology has made impressive steps forward in the creation of chimaeras of this sort, combining traits of living and non-living matter. Presently, cognitive robotics is the discipline that aims to design a synthetic autonomous being. In the last 30 years, autonomous robotics has attained increasingly complex objectives, as research on intrinsic motivation shows (Hester & Stone, 2017; Qureshi et al., 2018). In recent years, besides building machines capable of carrying out the most onerous and wearisome manual tasks, engineers have also designed robots capable of social interactions with humans. This enterprise has brought synthetic autonomous beings inside the realm of human sociality. Whilst some human-assisting devices – vacuum cleaners, machines moving heavy objects – require minimal interaction with humans, for many others interaction will be crucial to optimal functionality. Among current applications, we find virtual assistants, caring robots for children, elderly people, and people with disabilities. Among such interactive machines, it is possible to make a further distinction. On the one side, we have machines for which human interaction is an essential means to performing some further task but is not an objective in itself – virtual assistants, self-guided machines, robot waiters. On the other side, we find machines for which establishing an intersubjective relationship is itself the objective. These are the so-called companion robots, that group of (more or less) anthropomorphic machines that are designed and programmed specifically to produce an intersubjective interaction, often including an emotional dimension, with humans.

DOI: 10.4324/9781003459798-1

Examples include elderly care robots, robots that establish relationships with individuals on the autism spectrum, educational robots for children, and sex robots. In today's technological landscape, companion robots serve as a prime representation of human–machine interactions, encapsulating various interactional cues ranging from verbal communication to proxemics and facial expressions. However, while they showcase an impressive range of interaction methods, they may not always be the pinnacle of efficiency in the realm of social technologies. For instance, chatbots powered by large language models (LLMs) stand out as a form of social technology that, although primarily focused on verbal interaction, often deliver more streamlined social communication compared to some social robots (SRs). This book delves into the world of companion robots (CRs) and SRs, portraying them as paramount examples of technologies striving to mirror human social nuances. The discussions and arguments raised, while rooted in the context of these robots, have broader implications and can be adapted to encompass other interactive technologies. In the conclusion, we introduce the term "synthetic social agents" to describe not just social robotics but also a distinct approach for technologies to interface with the social system.

In fact, this book does not primarily delve into the philosophical implications of social robotics. Instead, its purpose is to offer a fresh perspective on the interplay between technology and social systems, highlighting social robotics as a prominent example. As readers will see, Chapters 3 and 4 touch on social robotics only tangentially. These two chapters are more interested in designing a theory able to capture the interplay between human and non-human agents shaping socio-technical systems. The central enquiry of this manuscript is not "what are the philosophical implications of social robotics" but rather "how can we construct a theory of social systems where both human and non-human agents stand on equal footing".

What is a social robot and how it interacts

How to allow a robot to interact socially with a human being? The notion of a robot's "sociality" is complex and subject to lively discussion – is the "Roomba" floor-cleaning robot social? It is certainly not trivial to understand, and then implement, the elements that allow human–robot interaction (HRI) to take place. This is a problem dealt with by so-called *interaction studies*, which enquire into the optimal strategies for interaction that allow a robot to be social. One thing that seems certain is that human beings, when confronted with an anthropomorphic robot, are inclined to behave accordingly (Krämer et al., 2011). It seems that the relevant elements for making an interaction social are, among others, facial expressions, gaze (Mutlu et al., 2009; Riek et al., 2010), proxemics (Dautenhahn et al., 2006; Mumm & Mutlu, 2011), non-verbal language in general (Bartneck et al., 2020b; Hegel et al., 2011; Knight, 2011), and – obviously – verbal language.

Some fundamental dimensions perceived by users as making an interaction "more social" are likeability, perceived trustworthiness, competence, adaptivity, embodiment, sociability, and agency. All these are typical items measured during experiments on HRI (Heerink et al., 2010) to understand which are the most effective interactional elements in human–robot relationships. A typical problem of HRI is to understand which interactional features influence these items and, above all, to understand the systemic relationship between them. A relationship, in fact, is not simply the result of combining multiple interactive and social features that, if brought to maximum efficiency, generate an effective human–robot relationship. The delicate balance between verbal and non-verbal contents, along with the interplay between verbal and non-verbal references and specifications, is of fundamental importance for the success of the interaction as well as to avoid humans' experiencing negative reactions or feelings of the uncanny (Kätsyri et al., 2015). Regarding the relevance for HRI of the systemic nature of human communication, we refer to another work where the issue of coherence between the robot's communicative registers is analysed in depth (Bisconti, 2021).

However, the literature on interaction studies is full of failures: robots that try to interact and are not even noticed (Satake et al., 2008), robots that generate anxiety through their gaze (Nomura & Kanda, 2015), and robots that annoy participants by their inability to step out of the "intimate zone" of proxemics (Dautenhahn et al., 2006). Starting from these failures, it has been possible gradually to identify the interactional features that, at least at the current state of the art, allow the design of SRs capable of relating successfully to humans.

Types of social robots

In the realm of technology, SRs have emerged as a fascinating intersection of artificial intelligence, robotics, and HRI (Breazeal, 2003). These machines, designed to engage with humans in a personal, emotive manner, are as diverse as the tasks they perform and the environments they inhabit (Dautenhahn, 2007). This section delves into the world of SRs, exploring their types, construction, interaction techniques, and historical development.

SRs are not a monolithic entity but rather a broad spectrum of machines, each with its unique technical capabilities, interactional nuances, and domains of use (Feil-Seifer & Mataric, 2011). They range from companion robots that provide emotional support, to service robots that assist with everyday tasks.

Consider Paro, a therapeutic robot resembling a cuddly baby seal (Wada & Shibata, 2007). Paro is more than just a charming companion; it's a sophisticated piece of technology designed to stimulate patients with dementia, Alzheimer's, and other cognitive disorders (Wang et al., 2022). Its ability to respond to touch, light, sound, temperature, and posture creates an illusion of emotional interaction, making it a valuable tool in healthcare settings. On the other end of the spectrum, we find humanoid robots like NAO. Developed by SoftBank Robotics, NAO is designed to mirror human behaviour and interaction. Its versatility makes it a popular choice in education, research, and healthcare. NAO can walk, talk, recognise faces and voices, and even detect obstacles in its path, making it a fascinating example of the potential of humanoid robots. Service robots like Tiago, developed by PAL Robotics, represent another facet of SRs (Pages et al., 2016). These robots are designed to assist humans in tasks such as serving food, carrying objects, or guiding people. Tiago, with its mobile base, torso, and arm, can navigate and manipulate objects in human environments, showcasing the practical applications of SRs.

The art and science of building social robots

Creating a SR is a complex process that involves several key components: sensing the environment, physical interaction, verbal and non-verbal interaction.

Sensing the environment is the robot's way of perceiving the world around it. For instance, NAO uses cameras for visual perception, microphones for auditory perception, and touch sensors for tactile perception. Similarly, Paro uses a combination of sensors, including light and temperature sensors, to interact with its environment. Physical interaction is another crucial aspect of SRs. Robots use actuators to move and interact with their surroundings. NAO, for instance, has 25 degrees of freedom, allowing it to walk, gesture, and manipulate objects. Tiago, with its extendable arm and gripper, can reach high shelves and carry objects, demonstrating the practical potential of physical interaction in SRs. Verbal and non-verbal interaction is what makes SRs truly "social". Robots like NAO can recognise and respond to human speech, making the interaction more natural and engaging.

Creating a SR that can interact smoothly with humans is a formidable task. It requires a delicate blend of various techniques from diverse fields such as machine learning (ML), computer vision, and natural language processing (NLP).

Below we describe some technologies that can be used to design SRs.

One of the most important, and advanced, abilities of a SR is to recognise emotion; this is a crucial aspect of social interaction (Breazeal, 2003). Emotion recognition technology enables robots to interpret human emotions by analysing facial expressions, body language, and tone of voice (McColl & Nejat, 2014). This allows the robot to respond appropriately to the user's emotional state, making the interaction more personal and engaging. For example, if a robot detects that a user is upset, it might respond with comforting words or actions. On the verbal side, NLP is often used in social robotics. NLP is a field of artificial intelligence that focuses on the interaction between computers and humans through natural language. It enables robots to understand and generate human language, making verbal interaction possible. With NLP, robots can understand commands, ask questions, and even engage in conversation with humans. This is crucial for creating a natural, engaging interaction between humans and robots. Very much connected with the ability of interacting verbally, SRs are usually equipped with speech synthesis, also known as text-to-speech (TTS). Speech synthesis is the process of generating spoken language from written text. This technology allows robots to communicate verbally with humans, making the interaction more natural and engaging. Modern TTS systems can generate speech that is almost indistinguishable from a human voice, with correct emphasis and intonation, even in different languages.

On the sensing sphere, we find computer vision. This is a field of artificial intelligence that trains computers to interpret and understand the visual world. In SRs, computer vision is used for tasks such as object recognition, activity recognition, and navigation. For instance, a robot might use computer vision to recognise a user's face, understand what the user is doing (like reading a book or watching TV), or navigate around a room without bumping into obstacles.

In general, many SR tasks are supported by ML. This is a type of artificial intelligence that enables computers to learn from and make decisions based on data. In SRs, ML can be used for a wide range of tasks, from recognising patterns in user behaviour to learning how to respond to different situations. For instance, a robot might use ML to learn how to recognise a user's preferences, habits, or emotions (Jain et al., 2019), allowing it to provide personalised assistance or recommendations. Specifically, reinforcement learning (RL) is often used to allow the robot to effectively learn effective interactions with its environment. RL is a type of ML where an agent learns to make decisions by taking actions in an environment to achieve a goal (Qureshi et al., 2018). The agent receives rewards or penalties for its actions, guiding it towards the most beneficial behaviour. In the context of SRs, RL can be used to train robots to perform tasks more effectively or to interact with humans in a more engaging manner. For instance, a robot might learn to recognise when a human is interested or bored during an interaction and adjust its behaviour accordingly.

The general field to which social robotics pertains is the one of HRI. HRI is a multidisciplinary field focused on understanding, designing, and evaluating robotic systems for use by or with humans (Bartneck et al., 2020a). It encompasses everything from the design of robot appearance and behaviour to the exploration of the social implications of robot deployment. In the context of SRs, HRI research helps to create robots that are safe, effective, and appealing to humans.

In conclusion, the creation of a SR that can interact smoothly with humans is a complex, multidisciplinary endeavour. It requires a blend of various techniques from fields such as ML, computer vision, and NLP. However, the result is a robot that can engage with humans in a personal, engaging, and effective manner, opening up exciting possibilities for the future of HRI.

A glimpse into the history of social robots and use cases

SRs have evolved to represent a compelling intersection between technology and human creativity. What was once confined to the realms of science fiction has, over time, been realised due to breakthroughs in both robotics and artificial intelligence.

The history of SRs can be traced back to the 20th century, with early attempts at creating machines that could interact with humans. However, it wasn't until the late 1990s and early 2000s that the field of social robotics truly began to take shape. This period saw the development of robots like Kismet, a robot developed at MIT that could recognise and respond to human emotions (Turkle, Breazeal, et al., 2006), and Paro, the therapeutic robot mentioned earlier. The development of SRs has been driven by a desire to create machines that can assist and interact with humans in a variety of contexts. From healthcare and education to customer service and home assistance, SRs have found a wide range of applications even if, up until now, major technical and interactional shortcomings have prevented them from an actual massive commercialisation.

Early experiments in social robotics, such as MIT's Kismet in the late 1990s, focused on non-verbal cues and emotion recognition (Turkle, Breazeal, et al., 2006), laying the groundwork for subsequent developments. This ushered in a new era of more complex robots like Honda's ASIMO (Shigemi et al., 2018), which incorporated human-like mobility and interaction. The introduction of Paro in 2003 marked a shift towards the therapeutic use of SRs. Its success demonstrated the potential for robots to establish emotional bonds with humans and offer psychological benefits, particularly for elderly or mentally ill patients.

In 2014, SoftBank's Pepper was introduced as one of the first robots designed to perceive and respond to human emotions, marking a significant advance in the field (Pandey & Gelin, 2018). Pepper's unveiling signalled the increasing importance of emotional intelligence in social robotics and opened the door to new possibilities for robot–human interaction. The last decade has witnessed an acceleration in the development of SRs, driven by advances in artificial intelligence and ML. Today's SRs can learn from their experiences, understand and respond to human language, and recognise and react to human emotions, blurring the line between technology and life. As the field of social robotics continues to evolve, so does its breadth of application. Here, we will explore five cases from scientific literature that highlight how SRs have been used across different settings.

Turkle's early work with robotic companions serves as a profound investigation into the emotional relationships between humans and SRs. In a study conducted in 2006 (Turkle, Taggart, et al., 2006), Turkle observed how people, especially children and the elderly, form emotional attachments to robotic pets like the Furby and the AIBO dog. She found that despite knowing these entities were simply machines, people often treated them with affection, care, and even empathy. Turkle's work has helped expose the intricate dynamics of HRI, raising essential questions about our growing reliance on technology for social and emotional needs.

NAO robots have been increasingly used as therapeutic tools for children with autism spectrum disorder (ASD) (Coeckelbergh et al., 2016). A study by Pennisi et al. (2016) showed that using NAO during therapy sessions could improve social interaction and communication skills among children with ASD. The child-friendly design and controlled, predictable interactions of NAO made it less threatening and more engaging for the children, making therapy sessions more productive. This study highlights the potential of SRs in therapeutic settings, providing new avenues for patient engagement and treatment.

Paro, the therapeutic seal robot, has seen extensive use in dementia care. A study by Jøranson et al. (2015) investigated the effects of interaction with Paro on mood, agitation, and social engagement in dementia patients. The results indicated that interaction with Paro led to increased social engagement and positive affect and decreased agitation among patients. This case underscores the role of SRs in healthcare, particularly in addressing the psychosocial needs of patients.

Robots like NAO have also found their place in classrooms, serving as innovative educational tools. In a study by Kory-Westlund and Breazeal (2019), a storytelling robot was used to promote children's learning and engagement. Children who interacted with the storytelling robot showed improved vocabulary retention and were more engaged in learning activities. This use case demonstrates how SRs can enhance educational experiences, fostering active learning and engagement.

Finally, SRs have been utilised in customer service settings. A study by Kanda et al. (2010) used a humanoid robot in a shopping mall in Korea to provide information and directions to shoppers. The study found that shoppers who interacted with the robot reported higher satisfaction levels and a more enjoyable shopping experience. This case illustrates how SRs can effectively interact with the public, potentially revolutionising customer service and retail environments.

These examples highlight the transformative potential of SRs across various domains, from healthcare and education to customer service. As our understanding and technology continue to advance, the applications of SRs are set to become even more diverse and impactful.

In recent years, the field of social robotics has continued to evolve, with advancements in artificial intelligence and ML leading to more sophisticated and capable robots. As these machines become more advanced and their interactions more nuanced, we may find ourselves living in a world where robots are not just tools but also integral parts of our social fabric.

Robots in society: ethics issues

In this regard, a major debate has arisen regarding the design and use of these interactive machines, still at a limited commercial stage, which are able to understand humans' words, gain their trust through caregiving, monitor their physical and psychological health, recognise emotions, simulate emotional states, and respond coherently to humans' verbal and non-verbal interactions. In addition to a general scepticism about the possibility of a positive public reaction to the introduction of such machines into everyday life (Coeckelbergh et al., 2016), other ethical concerns have been raised by researchers on this issue.

In the philosophical discussion, two topics have drawn most of the discussion's focus: the moral status of the machine and the so-called deception objection. On the first point, the discussion is so wide-ranging that it has become a separate field of research: *machine ethics*. Machine ethics is concerned with the moral status of machines in both an active and a passive sense, namely whether a machine can perform moral actions and whether morally relevant actions can be performed *on* a machine. Specifically, however, the greatest effort is certainly to provide the machine with a set of ethical principles and the ability to apply them – to allow it to act *morally* (Anderson & Anderson, 2011). This theoretical effort often revolves around case studies of moral dilemmas, such as the *trolley problem* for autonomous driving cars. There is also a second strand of research, which follows a more "continental" approach, identified with the concept of *roboethics*. If *machine ethics* and *roboethics* differ in anything, it is certainly the greater attention that machine ethics gives to the problem of the machine's morality, as opposed to the more socio-technical analysis characteristic of roboethics. The latter focuses on understanding how a given machine will modify the socio-technical system. The premise of roboethics is to consider the co-interaction, the *feedback loop*, that develops between the technological object and the social system and to enquire how the two elements co-construct themselves (Coeckelbergh, 2010). The technological object shapes society and is in turn shaped by it. Roboethics is the approach from which this work takes its cue as it considers the ethical issues raised by machines only within the social system. Our interest does not concern the machine *itself* but the machine as a cultural construct, as an object-signifier produced inside a certain social world. Therefore, the machine is *constructed* twice: once in the literal sense of being assembled and a second time in the way that its value is inextricably linked to the context in which it is created. Roboethics is, in short, much more interested in the social effect of the machine than in the machine itself.

The second issue widely addressed in the literature on social robotics is the deception objection. According to the scholars raising this objection, SRs are unethical because they may deceive the user about their non-human nature. In the words of Sparrow and Sparrow (2006),

> What most of us want out of life is to be loved and cared for, and to have friends and companions, not just to believe that we are loved and cared for, and to believe that we have friends and companions, when in fact these beliefs are false.

This assumption of a binary opposition between true and false companionship has already been the object of our criticism. We refer to Bisconti and Nardi (2018) to deepen the reader's doubts about such a binary view of relationships. Moreover, even in robotics for the elderly, it has not often (or even once, to our knowledge) been reported that subjects have actually been deceived by a robot with regard to its artificial nature. Even in Turkle's experiments, users like the elderly Andy, who experiences a strong *transfer* to the robot, remain aware of interacting with a machine (Turkle, Taggart, et al., 2006). In any case, the second part of this manuscript will attempt a correct framing of the problem of *deception*.

Other issues typically raised by social robotics include concerns related to privacy and user safety, especially in relation to social robotics for elders or children on the autism spectrum (Coeckelbergh et al., 2016).

Beyond these issues, the introduction of interactive robots into human society raises important questions of acceptability and social implications. Before deepening these issues, we must first make a step back with Mark

Coeckelbergh (2011a,b), who asks "How can we know and how should we evaluate the human–robot relationship?" This question brings the conversation back to a basic, fundamental level; we must start from here. This step back is necessary in order to clearly frame the gnoseological horizon of social robotics, so that ethical and value concerns can at least find a proper place. Before any attempt at an answer, however, Coeckelbergh's question requires some methodologically preliminary reflections. After that, in the final part of this introduction, we give a brief presentation of the objectives of this work.

Three preliminary considerations

How can we know and how should we evaluate the human–robot relationship?

(Coeckelbergh, 2011a,b)

To begin with, this question implies a distinction between two dimensions: the gnoseological and the axiological. The fact that the former must precede the latter is by no means to be taken for granted but is a peculiar characteristic of the specific theoretical object that is human–robot relations. This is due to a special "liminality" of these interactions, which are (dangerously) placed on the borderline between two extremely well-known and familiar models: intersubjective human relations and object relations. In fact, the robot is undoubtedly a "thing", an object created for and at the disposal of the human being: pure "thingness", from the Heideggerian *Gestell*, which inherits its name from "robota". This word meant "worker" in medieval Czech (Henry, 2009), while in Old Church Slavic it meant "servant, slave".[1] At the same time, a SR interacts, speaks, uses facial expressions and the non-verbal registers, almost seems to feel emotions and can certainly make humans feel them (Rosenthal-von der Pütten et al., 2014). Its style of interaction closely reproduces (and perhaps will one day perfectly reproduce) human interactional patterns and can establish something resembling an intersubjective interaction (Masaaki Kurosu, 2020). This is widely demonstrated in interaction studies: although we know a robot's artificial nature perfectly well, we tend to anthropomorphise it both in language (Coeckelbergh, 2011a) and in behavioural patterns (Krämer et al., 2011), including expectations of non-verbal patterns (Dautenhahn et al., 2006), even from a neurological point of view (Rosenthal-von der Pütten et al., 2014). How, then, can we truly know HRIs? Construing their patterns solely in terms of human interactions certainly fails to take some fundamental differences into account. As we have already addressed in another paper (Bisconti, 2021), subjects' awareness of a robot's machinic nature influences their interactions with it. But this is not the only issue: as an interactionist approach like that of Watzlawick (1971) reveals, the lack or inconsistency of verbal or non-verbal registers could create completely new interactional patterns. In short, we should enquire not only which are the elements that HRIs borrow from human–human interactions (HHI) but also and above all which peculiar interactional patterns can emerge from HRIs.

At the same time, the current design of SRs aims at imitating human interactions. The stunning similarity between human–human behavioural patterns and HRIs does suggest that the latter cannot be conceptualised as simply objectual (Richardson, 2016; Sharkey & Sharkey, 2006).

In short, a first challenge raised by social robotics is as follows: which categories should we use to think about human–robot relationships? Perhaps something is in between the intersubjective and the objectual, perhaps something completely different.

After that, of course, the axiological issue arises, which is no longer to describe but to *prescribe* the "right" setting for HRIs. This latter challenge is the focus of most of today's scientific literature, which attempts a conceptualisation of the "right" relationship with robots. Certainly, the lack of a clear theoretical framework to *describe* and *explain* HRIs does not help in this endeavour. Still, we are in the unusual situation of raising a normative issue about an object that does not precede our intervention. Usually, axiology aims to correct the Kantian "crooked wood", such as humankind, that is already present. A theory of "good" or "right" behaviour is usually preceded by the presence of "wrong" or "unfair" behaviour needing to be corrected. This does not apply to robots, which are designed and normed at the same time and do not pre-exist in their normative design. In this sense, *interaction design* in social robotics is definitively a political act: the simultaneity of the prescriptive and descriptive acts resembles creation *ex nihilo*, where the ontology of the robot is also an axiology, a precise choice of values. These suggestions will be further discussed in the following pages.

The final preliminary note to any discussion of social robotics, which introduces the eminently political aspect of the next sections of the book, concerns what we might call the "false consciousness" of the robot designer. Perhaps it is even banal, yet necessary because it is often forgotten, to recall that the normative effect of robot design, in its double action of describing and prescribing *ex nihilo* the onto-axiology of the robot, refers to the world of human beings. This means that when designers choose a specific haircut, perhaps long and feminine, for a robot receptionist, they are performing a prescriptive action not only for the robot but also for society itself. This goes beyond the concept of bias: the design of the robot, its body, is written and shaped by a socio-political competition between symbols, narratives, and mechanics of representational power. The robot is like a blank sheet of paper where symbolic structures and conflicting narratives can clash as in a proxy war – far from human bodies, yet still referring to them. The topic of the robot as a "place of alienated symbolic production" will be deepened in this work, both in its political implications and in psychological considerations on the hallucinatory nature of HRIs.

Objectives of this work

The aim of this work is to discuss and understand from both a socio-political and a subjective-relational point of view the human–machine hybridisation of social systems. The introduction of social robotics raises the hybridisation and contamination between humans and non-humans to a new level. Up until now, the dimension of sociality has still largely belonged to humans, with technologies acting in the background. Even the use of artificial intelligence as a means of regulating and generating social spaces, such as social networks, did not imply a direct hybridisation of human sociality as these technologies only acted as a facilitating medium and not as social actors themselves. Although they structurally modified human social practices, these technologies still worked as *media*. With social robotics, on the other hand, it is necessary to rethink the interactional and symbolic space of sociality from a perspective that welcomes hybridisation processes not as an "addition" to the interactions between human beings but as a co-essential element. This means changing the focus of hybridisation from the individual hybrid subject, such as a cyborg, to hybrid socio-symbolic spaces populated simultaneously by humans and non-humans. Such a perspective requires an overturning of the anthropocentric viewpoint, which is essentially based on the assumption that the only social actors are human beings.

This work unfolds on three levels, each of which builds on the last. The first is a socio-political analysis of social robotics, to correctly relocate the relationship between the robot and the social. In fact, as our brief analysis with Coeckelbergh has shown, the problem that logically precedes any other is a correct descriptive theory for a sociality hybridised by interactive machines, whose specificity and peculiarity we have briefly sketched. This re-location will challenge the liminal zone between human and non-human in the anthropocentric tradition. The scope of the first part of this manuscript is to understand under what conditions a technology, such as SRs, will change society as a whole. This question presupposes a theory of the functioning of socio-technical systems; the fourth chapter will make an attempt in this direction. To begin with, we will reshape our understanding of the concept of "social system" through Bruno Latour's actor–network theory, the concept of "sociomorphing" from Johanna Seibt et al.'s (2020), and through Federica Russo's (2022) concept of poietic agent. We will show

how non-human actors can be considered on the same level as human ones in shaping the social. Then, we will face an issue that will require stepping back from social robotics to build a theory of social system crisis. This will lead us into systems theory applied to the social, through the lenses of Humberto Maturana and Francisco Varela and the contributions of Gilbert Simondon and Alain Badiou. Finally, we will apply our conclusion to the analysis of the political implications of SRs.

The second level undertakes an analysis of the nature of one-to-one HRI in order to understand the implications of HRI at the level of subjective and intersubjective organisations. Sherry Turkle et al. (2006), Davide Gunkel (2017), and Coeckelbergh (2011a,b) have already underlined the difficulty of conceptualising the human–robot relationship from a psychological and philosophical point of view. In this second section, we will attempt to answer two questions: what, from the point of view of relational mechanisms, is the unique nature of human–robot intersubjective relations? How can they influence, alter, or contaminate human–human relations, and with what effects? In this discussion, we will employ different perspectives, ranging from the current discussion of the "deception objection" (Sparrow & Sparrow, 2006) to the framework of the "symbolic argument". After reviewing existing approaches, we will introduce a theoretical solution addressing the relational implications of HRI, drawing on conceptual tools from Donald Winnicott, Jessica Benjamin, and Jacques Lacan. This book owes much to Lacan, whose ideas serve as a foundational pillar throughout. Notably, the French psychoanalyst plays a crucial role in challenging and dissecting the relationship between the subject and the external world – a key focus in Chapters 4 and 6. From a theoretical standpoint, this book aims to connect Lacan's theory of the Real with the systemic approach of Maturana and Varela, illustrating how these two frameworks can converse and complement each other. This endeavour may be contentious for various readers and for a range of reasons. Devotees of Lacan might resist the idea of reconciling his work with systemic theory. Additionally, many philosophers of technology might be sceptical of integrating French philosophy and psychoanalysis into their field, viewing such an approach as too abstract or nebulous. However, I believe the subsequent pages, especially between Sections "Networks of meanings" and "The crisis of socio-symbolic systems", will elucidate the invaluable insights Jacques Lacan offers. Specifically, when read in a particular light, Lacan becomes instrumental for two primary objectives. Firstly, he provides a foundational gnoseological perspective on the intricate relationship between a system and its environment, especially when formulating a theory of crisis for socio-technical systems. Through Lacan, we gain clarity on the enigma: "How do social systems restore stability after a crisis?" Secondly, his notion of the "object petit a" offers a more profound comprehension of the term "otherness" in the context of human–robot relations. While some readers might view choosing Lacan as an unconventional decision, I encourage an open-minded exploration of this audacious attempt.

In the final sections, we will discuss some implications of this book for the design of SRs, sketching three conclusions of a non-anthropocentric view of hybrid socio-technical systems: (i) sociality is a property of a system; (ii) anthropomorphic interactions are, and always have been, a subclass of sociomorphic interactions; (iii) analysing hybrid interactional systems reveals new forms of social acting.

This work does not faithfully follow a specific conceptual strand but owes its approach to a multiplicity of theoretical suggestions. On the analysis of the relationship between robots and societies, the frameworks of actor–network theory and Maturana's autopoiesis intersect in the problematisation of anthropocentrism. In the investigation of the psychological universe of HRI, the main references will be Winnicott's psychoanalysis, Benjamin's recognition theory, and conceptual suggestions from Lacan's concept of phantasmatic support. All these conceptual models will converge in the effort to reconceptualise the relationships between robot and society and between robots and humans. In the concluding sections, we depict a research agenda for correctly enquiring into hybrid social systems.

As many academics are keenly aware, the time allocated for reading papers and books can be limited. Sometimes, the effort expended on a lengthy book doesn't justify its perceived value, and many times the insights for one's research were contained in just few paragraphs. Therefore, before we embark on this journey, here's a brief guide for readers.

If you are primarily interested in an actor–network approach to social robotics, focus on Sections "ANT: flattening the social" and "Who is enrolling whom". Afterwards, proceed directly to Chapter 7, where you'll discover a fresh perspective on the non-anthropomorphic elements of HRIs. For those solely concerned with the discussion on the implications of HRI for human relationality, Chapter 6 will be of interest. Here, we present a theory exploring the conditions under which relational patterns in HRIs might influence a user's relational dynamics. If you're keen on a novel theory of socio-technical systems, probably the most interesting and innovative theoretical contribution of this manuscript, it would be best to engage with the entire book. However, if you're already familiar with social robotics or find the specifics of SR technology not relevant for you, Chapter 4 is key. Within that chapter, I attempt to elucidate the circumstances leading to the collapse and rebirth of social systems. Subsequently, Chapter 7 offers insights linking back to the philosophy of technology.

Note

1 www.etymonline.com/word/robot via wordnet (consulted 18/06/2021).

References

Anderson, M., & Anderson, S. L. (2011). *Machine ethics*. Cambridge University Press.

Bartneck, C., Belpaeme, T., Eyssel, F., Kanda, T., Keijsers, M., & Šabanović, S. (2020a). *Human-robot interaction: An introduction*. Cambridge University Press.

Bartneck, C., Belpaeme, T., Eyssel, F., Kanda, T., Keijsers, M., & Šabanović, S. (2020b). Nonverbal Interaction. *Human-Robot Interaction, 9781108735,* 81–97. https://doi.org/10.1017/9781108676649.006

Bisconti, P. (2021). How robots' unintentional metacommunication affects human–robot interactions. A systemic approach. *Minds and Machines, 31*(4), 487–504. https://doi.org/10.1007/s11023-021-09584-5

Bisconti Lucidi, P., & Nardi, D. (2018). Companion robots: The hallucinatory danger of human-robot interactions. *AIES 2018 – Proceedings of the 2018 AAAI/ACM Conference on AI, Ethics, and Society,* 17–22. https://doi.org/10.1145/3278721.3278741

Breazeal, C. (2003). Emotion and sociable humanoid robots. *International Journal of Human-Computer Studies, 59*(1–2), 119–155.

Coeckelbergh, M. (2010). Robot rights? Towards a social-relational justification of moral consideration. *Ethics and Information Technology, 12*(3), 209–221. https://doi.org/10.1007/s10676-010-9235-5

Coeckelbergh, M. (2011a). You, robot: On the linguistic construction of artificial others. *AI and Society, 26*(1), 61–69. https://doi.org/10.1007/s00146-010-0289-z

Coeckelbergh, M. (2011b). You, robot: On the linguistic construction of artificial others. *AI and Society, 26*(1), 61–69. https://doi.org/10.1007/s00146-010-0289-z

Coeckelbergh, M., Pop, C., Simut, R., Peca, A., Pintea, S., David, D., & Vanderborght, B. (2016). A survey of expectations about the role of robots in robot-assisted therapy for children with ASD: Ethical acceptability, trust, sociability, appearance, and attachment. *Science and Engineering Ethics, 22*(1), 47–65.

Dautenhahn, K. (2007). Socially intelligent robots: Dimensions of human–robot interaction. *Philosophical Transactions of the Royal Society B: Biological Sciences, 362*(1480), 679–704.

Dautenhahn, K., Walters, M., Woods, S., Koay, K. L., Nehaniv, C. L., Sisbot, A., Alami, R., & Siméon, T. (2006). How may I serve you? A robot companion approaching a seated person in a helping context. *Proceeding of the 1st ACM SIGCHI/SIGART Conference on Human-Robot Interaction – HRI '06, April 2005,* 172. https://doi.org/10.1145/1121241.1121272

Feil-Seifer, D., & Mataric, M. (2011). Socially assistive robotics. *IEEE Robotics & Automation Magazine, 18*(1), 24–31. https://doi.org/10.1109/MRA.2010.940150

Gunkel, D. (2017). The Changing Face of Alterity. Rowman & Littlefield.

Heerink, M., Kröse, B., Evers, V., & Wielinga, B. (2010). Assessing acceptance of assistive social agent technology by older adults: The Almere model. *International Journal of Social Robotics, 2*(4), 361–375. https://doi.org/10.1007/s12369-010-0068-5

Hegel, F., Gieselmann, S., Peters, A., Holthaus, P., & Wrede, B. (2011). Towards a typology of meaningful signals and cues in social robotics. *Proceedings – IEEE International Workshop on Robot and Human Interactive Communication*, 72–78. https://doi.org/10.1109/ROMAN.2011.6005246

Henry, B. (2009). Immaginario, culture e identità artificiali. The myth of Cyborgs. *Cosmopolis, 1*, 167–178.

Hester, T., & Stone, P. (2017). Intrinsically motivated model learning for developing curious robots. *Artificial Intelligence, 247*, 170–186. https://doi.org/10.1016/j.artint.2015.05.002

Jain, D. K., Shamsolmoali, P., & Sehdev, P. (2019). Extended deep neural network for facial emotion recognition. Pattern Recognition Letters, 120. pp. 69–74.

Jøranson, N., Pedersen, I., Rokstad, A. M. M., & Ihlebaek, C. (2015). Effects on symptoms of agitation and depression in persons with dementia participating in robot-assisted activity: A cluster-randomized controlled trial. *Journal of the American Medical Directors Association, 16*(10), 867–873.

Kanda, T., Shiomi, M., Miyashita, Z., Ishiguro, H., & Hagita, N. (2010). A communication robot in a shopping mall. *IEEE Transactions on Robotics, 26*(5), 897–913.

Kätsyri, J., Förger, K., Mäkäräinen, M., & Takala, T. (2015). A review of empirical evidence on different uncanny valley hypotheses: Support for perceptual mismatch as one road to the valley of eeriness. *Frontiers in Psychology, 6*(March), 1–16. https://doi.org/10.3389/fpsyg.2015.00390

Knight, H. (2011). Eight lessons learned about non-verbal interactions through robot theater. In B. Mutlu (Ed.), *ICSR 2011, LNAI 7072* (LNCS, pp. 42–51). Springer. https://doi.org/10.1007/978-3-642-25504-5_5

Kory-Westlund, J. M., & Breazeal, C. (2019). A long-term study of young children's rapport, social emulation, and language learning with a peer-like robot playmate in preschool. *Frontiers in Robotics and AI, 6*, 81.

Krämer, N. C., Eimler, S., von der Pütten, A., & Payr, S. (2011). Theory of companions: What can theoretical models contribute to applications and understanding of human-robot interaction? *Applied Artificial Intelligence, 25*(6), 474–502. https://doi.org/10.1080/08839514.2011.587153

Kurosu, Masaaki. (2020). Human-computer interaction multimodal and natural interaction PART2. In *22nd International Conference, HCII 2020*.

McColl, D., & Nejat, G. (2014). Recognizing emotional body language displayed by a human-like social robot. *International Journal of Social Robotics, 6*(2), 261–280. https://doi.org/10.1007/s12369-013-0226-7

Mumm, J., & Mutlu, B. (2011). Human-robot proxemics: Physical and psychological distancing in human-robot interaction. *HRI 2011 – Proceedings of the 6th ACM/IEEE International Conference on Human-Robot Interaction*, 331–338. https://doi.org/10.1145/1957656.1957786

Mutlu, B., Yamaoka, F., Kanda, T., Ishiguro, H., & Hagita, N. (2009). Nonverbal leakage in robots: Communication of intentions through seemingly unintentional behavior. *Proceedings of the 4th ACM/IEEE International Conference on Human Robot Interaction – HRI '09*, 2(1), 69. https://doi.org/10.1145/1514 095.1514110

Nomura, T., & Kanda, T. (2015). Influences of evaluation and gaze from a robot and humans' fear of negative evaluation on their preferences of the robot. *International Journal of Social Robotics*, 7(2), 155–164. https://doi.org/ 10.1007/s12369-014-0270-y

Pages, J., Marchionni, L., & Ferro, F. (2016). Tiago: The modular robot that adapts to different research needs. *International Workshop on Robot Modularity, IROS, 290.*

Pandey, A. K., & Gelin, R. (2018). A mass-produced sociable humanoid robot: Pepper: The first machine of its kind. *IEEE Robotics & Automation Magazine*, 25(3), 40–48.

Pennisi, P., Tonacci, A., Tartarisco, G., Billeci, L., Ruta, L., Gangemi, S., & Pioggia, G. (2016). Autism and social robotics: A systematic review. *Autism Research*, 9(2), 165–183.

Qureshi, A. H., Nakamura, Y., Yoshikawa, Y., & Ishiguro, H. (2018). Intrinsically motivated reinforcement learning for human–robot interaction in the real-world. *Neural Networks*, 107, 23–33. https://doi.org/10.1016/j.neunet.2018.03.014

Richardson, K. (2016). Technological animism: The uncanny personhood of humanoid machines. *Social Analysis*, 60(1), 110–128. https://doi.org/10.3167/ sa.2016.600108

Riek, L. D., Paul, P. C., & Robinson, P. (2010). When my robot smiles at me: Enabling human-robot rapport via real-time head gesture mimicry. *Journal on Multimodal User Interfaces*, 3(1–2), 99–108. https://doi.org/10.1007/s12 193-009-0028-2

Rosenthal-von der Pütten, A. M., Schulte, F. P., Eimler, S. C., Sobieraj, S., Hoffmann, L., Maderwald, S., Brand, M., & Krämer, N. C. (2014). Investigations on empathy towards humans and robots using fMRI. *Computers in Human Behavior*, 33, 201–212. https://doi.org/10.1016/j.chb.2014.01.004

Russo, F. (2022). *Techno-scientific practices: An informational approach.* Rowman & Littlefield.

Satake, S., Kanda, T., Glas, D. F., Imai, M., Ishiguro, H., & Hagita, N. (2008). How to approach humans? Strategies for social robots to initiate interaction. *Proceedings of the 4th ACM/IEEE International Conference on Human-Robot Interaction, HRI'09*, 109–116. https://doi.org/10.1145/1514095.1514117

Seibt, J., Vestergaard, C., & Damholdt, M. F. (2020). Sociomorphing, not anthropomorphizing: Towards a typology of experienced sociality. In M. Nørskov, J. Seibt, O. S. Quick, *Culturally sustainable social robotics*. Frontiers in Artificial Intelligence and Applications (Vol. 335, pp. 51–67). IOS Press. https://doi.org/ 10.3233/FAIA200900

Sharkey, N., & Sharkey, A. (2006). Artificial intelligence and natural magic. *Artificial Intelligence Review*, 25(1–2), 9–19. https://doi.org/10.1007/s10 462-007-9048-z

Shigemi, S., Goswami, A., & Vadakkepat, P. (2018). ASIMO and humanoid robot research at Honda. In *Humanoid robotics: A reference* (pp. 55–90). Springer.

Sparrow, R., & Sparrow, L. (2006). In the hands of machines? The future of aged care. *Minds and Machines*, *16*(2), 141–161. https://doi.org/10.1007/s11 023-006-9030-6

Turkle, S., Breazeal, C., Dasté, O., & Scassellati, B. (2006). Encounters with kismet and cog: Children respond to relational artifacts. *Digital Media: Transformations in Human Communication*, *120*, 1–20.

Turkle, S., Taggart, W., Kidd, C. D., & Dasté, O. (2006). Relational artifacts with children and elders: The complexities of cybercompanionship. *Connection Science*, *18*(4), 347–361. https://doi.org/10.1080/09540090600868912

Wada, K., & Shibata, T. (2007). Living with seal robots – Its sociopsychological and physiological influences on the elderly at a care house. *IEEE Transactions on Robotics*, *23*(5), 972–980.

Wang, X., Shen, J., & Chen, Q. (2022). How PARO can help older people in elderly care facilities: A systematic review of RCT. *International Journal of Nursing Knowledge*, *33*(1), 29–39.

Watzlawick, P. (1971). *Pragmatica della comunicazione umana*. Astrolabio.

2 Beyond subjects and objects

Actors in hybrid societies

The human social world has always been crowded with talking objects, fantastic beings, and non-human helpers who intervene intentionally in the physical world: the river spirit of an animist religion, the magic lamp of Middle Eastern fairy tales. For every culture and every human being's imagination, objects have never simply been still, inert, and silent. This "animacy" of matter can perhaps be traced back to an ancestral form of anguish about the unpredictability of nature, as Feuerbach argues when discussing the birth of religion (Feuerbach, 1972). The human being is alone before a world that is not friendly, not even an enemy, but simply indifferent. The external object is not only threatening, like lightning or fire; above all, it is incomprehensible, highlighting the gnoseological deficiency that so distresses human beings. Perhaps it is for this reason that humans animate the world – to make it controllable. And in the grip of a neurotic impetus, as if to placate this anguish once and for all, humans create the machine. Creation ex nihilo, that reason for envying God, is not only the reduction of reality to the imagination; above all, it is the capacity to fill the chasm that divides the gnoseology of subjects from the ontology of *entia*, in order to definitively obliterate the structural unpredictability of reality. As Ernst's spinning-wheel game (Freud, 1955) is a formal, compulsive scheme to repeat a specific repressed content, the repeated ex-nihilo creation of machines is evidence of an archetype: the desire to make fantasy and reality coincide in order to extinguish the gnoseological anguish due to an unpredictable and unreliable world. If, following Feuerbach, humans alienate themselves from the ability to bridge the gap between ontology and gnoseology by ascribing it to God, they reappropriate that ability in the creation of the machine. They transform the unpredictable and dangerous *ens* into an *object* that can be known and controlled. In this sense, the talking machine is perhaps an archetypal idea. Is it therefore under the frame of epistemic violence that humans construct the relationships they have with the external world? There is no doubt that a long and flourishing

DOI: 10.4324/9781003459798-2

philosophical tradition considers the problem of object relations in these terms: the process of objectification is a domestication aimed at harnessing, categorising, and normalising the otherness of the *ens* in the familiarity of the object. When operative in human intersubjective relationships, this process has a derogatory element. The process of recognition is structurally linked to the "work of the negative", which in Hegelian phenomenology requires a moment of misrecognition between subjects; at a later moment recognition is possible, but only between human beings. Why? Firstly, because human matter resists objectification, a concept metaphorically reformulated as the "struggle for life and death". And yet, nonhuman matter also resists, perhaps even harder at times: how much does raw gold dug from the mine resist the objectification of becoming a ring? How much energy is required to extract it, transport it, bend it to the shape of the finger? Perhaps not much less than it takes for a human to be bent to the will of another: trivially, a human can be persuaded, whereas gold cannot. And yet, self-consciousness in its first moment goes about a world "at its disposal" denying its autonomy, devouring the *ens* and transforming it into an object in a sort of gnoseological-digestive process – and its omnipotence is undisturbed: the *ens* is docile, the otherness malleable. The difference might perhaps lie in the inertia of the *ens* in comparison to the struggle between humans. At most, the *ens* exhibits a form of passive resistance, the persistence of matter, rather than an active clash of self-consciousnesses. Therefore, humans do not risk being objectified in turn by an object; at most, they might fail in objectifying the *ens*. Yet even this frame does not seem to be close to the experience of subjects. Feuerbach himself argues that the natural world, the world of *entia*, is so frightening and deadly that human beings animate and anthropomorphise it in order to curry favour with prayers and sacrifices. In short, the "objects" seem to be less inert, less stable, and harmless than they have been depicted as being. If the natural *entia* retain a terrifying, distressing, and potentially fatal aspect, could the object of anthropic experience, already normed in gnoseological categories, be completely subjugated to the human? Perhaps, as Husserl (1968) proposed with "categorial intuition", the inkwell could have its own independent "inkiness", and thus the otherness of the object, even when anthropic, could not be reduced to nothing. For a long time, Western thought relegated the role of *entia* to objects, their action to their function, and their reality to omnipotent fantasy.

If the Anthropocene has long since begun, it is certainly characterised not by the action of the individual human subject but by that of the human collective. In order to shape the world around them, humans coordinate themselves in a dense network of linguistic, symbolic, and objectual transactions that constitute the social. Attempts to understand the objectual nature of the social, and conversely the social nature of the objectual, have so far been rare and largely ignored. Yet, the introduction of social robotics into our societies raises the urgency of this issue above its past levels. The robot, the object created ex nihilo by anthropic efforts, could turn out to be much more "*ens*" and "other" than expected. It might upset the overriding status of the human subject that maintains the difference between objectual and intersubjective relations; it might subvert the exceptional status of the human. If, therefore, to make an object it is necessary to mute and tame an *ens* with a disciplining gnoseology, perhaps in making an *ens* we should let the object speak. However, the task is the opposite of what it once was. The old problem humans faced was to emerge from the undifferentiated matter that constituted the real and construct intersubjectivity as a structure of differentiation. Now, however, we face the danger of a newfound fusion and confusion with the world of the *entia*, which this time comes from the objects approaching the threshold of intersubjectivity, identifying and delimiting the human with respect to nature. In short, if a robot can interact socially in the same way as a human, what basis is left for humans standing out from *entia*? This intermingling, this contamination, breaks the boundaries between those who had to be "on stage", the actors, and those who had to remain in the audience, the objects. The social robot interacts, watches with curious eyes (Mutlu et al., 2009), and metacommunicates like a Watzlawick patient (Watzlawick, 1971). It builds emotional bonds with humans (Rosenthal-von der Pütten et al., 2014). The items on a Likert-scale questionnaire about interactions with a social robot include the robot's *trustworthiness, likeability, competence*, and *sociability* (Heerink et al., 2010) – all attributes plundered from the subject-human to describe the object-robot. This mixture is inauspicious for anthropocentrism, which is rapidly crumbling before a world in which humans and non-humans easily exchange functions and principles of action. It may seem paradoxical, but the ex-nihilo creation of the social machine by the omnipotent human being, a supreme act of gnoseological ὕβρις, achieves exactly the opposite of the desired result. Instead of definitively obliterating anguish through the objectification of *entia*, it reveals the structural instability of the ideological construction that distinguishes the human from the non-human, the interactive from the inert, the socially significant from the insignificant. So perhaps we need to reformulate the question: is it really *necessary* that the human relationship with objects be based on gnoseological violence? Far from being merely descriptive, the answer to this question prescribes the relationship between subjects and objects (Henry, 2013). The very act of asking this question already undermines the unstable roots of the object–subject distinction, opening the way to an a-anthropocentric approach.

Steps to reaching a hybrid social

With these considerations in mind, this work has a number of objectives, which will be presented as the discussion allows new issues and questions to emerge. Firstly, we ask: what theory can frame a non-anthropocentric understanding of societies hybridised by the non-human? This calls for us to investigate how to conceptualise a society where the relationship between objects and subjects is no longer structured in a hypostatised and surreptitious anthropocentrism. In this sense, we state from the outset that any sociological or socio-political theory is far from being merely a description, as if it were a neutral operation of annotating the players on the field. On the contrary, it is a strongly prescriptive, and therefore politically relevant, mark impressed on reality. Note the double nature of the verb "to order", meaning, on the one hand, the operation of putting things in the right places, and on the other, the operation of prescribing the right places for things. If, therefore, every theory of the social is an intimately political act, we consciously perform this act by embracing an anti-anthropocentric position in order to seek, through this act of ordering, an a-anthropocentric one. We find here an old problem encountered by second-wave feminism: how to escape the words pre-constituted by the previous ideological order – for Luce Irigaray, the phallogocentric order – without being either pro- or "anti", instead becoming "a-". If illogic is the negation of logic, the a-logical is a moment prior or subsequent to logic itself. In the same way, anti-anthropocentrism must be seen as a bridge-passage towards a new modality, an "a-anthropocentric" one. To escape the dichotomy of being already described by the symbolic order, in either the form of participation or the form of refusal, is the typical problem of any theory purporting to be revolutionary. This work does not set the unattainable goal of making this transition, yet we hope at least to lay the foundations of its thinkability.

Secondly, in the next chapters we will raise the problem of crisis: what are the conditions that, within a given "social" context, provoke a profound and essential restructuring of the forces in the social field and, ultimately, of the symbolic structure of societies? In order to know whether social robotics will have a displacing, contaminating, and disrupting effect, we must be able to understand what the mechanics and conditions are that can give rise to a restructuring of the social paradigm. The introduction of objects, of natural or human forces, of new modes of socio-economic relations, changes history in a way that is difficult to predict. Trivial examples of this difficulty are the crises of social systems: why do they occur at certain moments in history and not others? Why are they triggered by certain specific events and not others? At least for an instant and with an unrepeatable synchrony, objects, forces, relationships, and the socialised experiences of subjects seem to cooperate to give unity to an apparently stochastic movement, history. So how does it happen that a specific object or process acquires a significant role within this particular, quasi-chemical precipitation of elements? Certainly, the overused concept of "emergent phenomenon of complex systems" would be a good way to give a conceptually appealing formulation to something that is not understood. Yet, we believe that on the phylogenetic profile, a theory of crisis should consider what Maturana suggested about the structure of autopoietic units. This approach, in combination with those of Latour and Gilbert Simondon, seems to apply particularly well to the uncoordinated occurrences of crises in human societies. We claim that a theory of crisis that applies to hybrid human/non-human social systems will allow us to understand how, and under what conditions, artificial social agents (and on top of them, social robots) can actually modify the structure of societies. In order to understand the displacing effects of animate and talking objects, it is therefore necessary to understand under what conditions and according to what mechanisms the crises of symbolic systems are generated.

The limitations of the ontological approach to object agency

We remain faithful to the idea that "theory" is not something neutral but rather what orders this world. Therefore, we aim to develop an approach overcoming the dualistic and almost Manichean view of the world that strongly differentiates between objects and subjects. The previous discussion led us down the path of the Hegelian concept of self-consciousness, because (to begin with) the social is traditionally divided into those who act and those who undergo, incurring the other's action.

The concept of *agency* represents the fundamental political-metaphysical frame that produces the distinction between subjects and objects. The subject is an agent, while the object is exposed to others' actions and is therefore malleable by gnoseological manipulations. Intentional actions are at the core of the distinction that safeguards humans from falling back into the magma of *entia*. Actions should have some characteristic by which they may be recognised as such. For if actions required only having some effects in the world, one could hardly argue that an earthquake does not *act*. The action of human subjects, on the other hand, is said to be essentially different: it is not subject to the strict mechanistic laws of material reality. This difference, though not detectable phenomenologically, is embodied in a multiplicity of concepts, such as free will, intentionality, and agency, with all the conceptual differences and overlaps entailed in these concepts. The issue of agency has also concerned the artificial intelligence and robotics communities for a long time, where discussion on this topic was quick to arise.

A well-known and widely discussed example is John Searle's "Chinese Room" (Searle, 1980). The target of Searle's argument was the view that mental processes are computational processes on formally defined elements, a claim attributed mainly to the supporters of so-called "strong AI". That line of thought on AI supported the view that a computational machine, which performs formal manipulation of symbols, not only is able to behave in the same way as a human being but actually has cognitive states and the ability to *understand*. We will not dig into Searle's argument, which the reader probably knows, but will skip to the conclusion of his thought experiment: we cannot state that the behaviour of a machine, even if indistinguishable from that of a human through phenomenological observation, implies agency. In the specific example discussed by Searle, *agency* is closely related to the ability to have semantic understanding – ultimately, intentionality. Given a virtually infinite amount of time and a very high computational capacity, the syntactic abilities of a machine can behaviourally *simulate* semantic understanding; Searle wants us to question whether that is sufficient for *actual* understanding.

This argument raises a striking challenge to the proponents of strong AI and thus to the functionalist approach. This last can be summarised as follows: the mind is a program that runs on hardware called "the brain" and can be implemented on virtually any other hardware that is equally powerful; at most, there might be "interoperability" issues. In contrast, Searle's account of intentionality appeals to a certain sort of "causal power", a characteristic he attributes exclusively to biological systems, making the *pars construens of* his argument far less solid than the *destruens*.

As the *querelle* in the Chinese Room shows clearly, setting the problem of agency in terms of the ontological differentiation of what an action is and what the attributes of agency are quickly shows the limitations of that approach.

For the same reason, we believe that the interest that the emerging *Object-Oriented Ontology* (hereafter OOO) is generating may lead only a little further towards a non-anthropocentric understanding of social systems hybridised to include non-humans. In spite of the fascination that the Heideggerian suggestions of OOO can provoke, it seems that this new "opaque" ontology, as many have called it, takes quite a reactionary posture towards the problem of "gnoseological violence" exposed in the previous chapter. We can summarise the positions of the OOO in the following points:

1 It insists on a kind of structural "interiority" of the object that is far from being actually known.
2 It considers all kinds of objects, including of course human beings, to be on the same ontological level – a position that would allow OOO to define itself as anti-anthropocentric.

3 It insists on the ontological symmetry of human–human and object–object relations, as well as human–object relations.

4 It claims that the "objectness" of the object is neither reducible to its relations, nor to its qualities, nor to any other phenomenal/relational/systemic structure. Ultimately, a kind of structural otherness must be preserved, prioritising "pure" ontology over any form of gnoseology.

For a general discussion of OOO, we refer to Graham Harman, the forerunner of this line of thought (Harman, 2018). In this work, we will not go into a detailed critique of the arguments brought by the OOO, for which we refer to some texts that meticulously underline the inconsistencies (Boysen, 2018; Gevorkyan & Segovia, 2020; Morton, 2011; Žižek, 2016).

We simply argue that, from a political perspective, OOO constitutes a dangerous shift from analysis of the practices and dynamics of domination produced by anthropocentrism to a kind of contemplative–spiritual retreat. From the point of view of political philosophy, we do not want to point out only the abstract gnoseological nature of these domination practices. In fact, anthropocentrism is at the core of the socio-political practices and narratives that structure the power relations between subjects. At first glance, OOO seems to be an excellent ally of the current ecological philosophies attempting to conceptualise a being-in-the-world that is not predatory and no longer destructive. In our opinion, however, it actually defuses all their revolutionary potential. If what concerns OOO about objects is their elusive ontological structure and not their fate in the world, it shifts the focus away from the political. The "mundane" and worldly fate of objects is relegated to a mere non-substantial accident of their only relevant aspect, the ontology of *entia*, remaining elusive, untraceable, and structurally dislocated. The exploitation of a tract of land is then no longer a problem to deal with since, in its very *gestalt*, this act does not cause any harm or ultimately affect the being of the *ens*, for the act is absent from the "worldliness" of the object. This somewhat cunning move to a-analytically flatten the problem of ontology – Harman (2015) makes a clarification that we believe is not significant here – onto the "opaque" plane of withdrawing objects fails to take responsibility for the political challenges that ontology inevitably entails. Like a modern night where all cows are black, Harman's ontology recalls Friedrich Schelling's undifferentiated absolute. What kind of analysis of the world can result from such "ineffability" of the object? This modern Damascius-like approach encompasses dangerous political–philosophical drifts, typical of a certain Heideggerian approach to the issue of technology. For if technology, and the socio-technical systems themselves, are non-essential, they also become theoretically irrelevant. The analysis of the contradictions of reality loses its meaning, along with any possible political philosophy. We must there-fore, in our opinion, beware of such approaches that, in order to solve a problem, simply stop looking at it.

In order to remain loyal to a political theory of social systems hybridised by non-humans, we believe it is necessary to abandon any attempt at defining agential ontological priorities and to land instead on a political understanding of action. We do not claim that the ontological question itself is not worthy of elaboration. However, we believe that grounding the understanding of the social and its actors in ontological differences in the attributes of agency is in itself an ideological choice. This choice is not neutral from the point of view of what we might call "gnoseological pol-itics": to ground the problem of citizenship in the social world in an onto-logical foundation is an excluding posture and reaffirms an anthropocentric stance. Along these lines, we believe there has been a misunderstanding of Latour's actor-network theory (henceforth ANT). Indeed, scholars have pointed out a criticism of ANT similar to the criticism of OOO just offered.

In his book *Reassembling the Social* (Latour, 2005) and in many other writings, Latour actually uses the term *agency* in association with objects. At first glance, this might seem like another attempt to attribute to objects the characteristics of human agency. On the contrary, however, it would be more adherent to the aims and methods of ANT to discuss the diffe-rence between *intermediaries* and *mediators*. It is through attention to this difference that we hope to dispel some of the fog that hangs over hybrid systems. In the next chapter we will introduce the main concepts of ANT, highlighting the aspects that are particularly relevant to our work. In the following chapter we will show how the study of social robotics encounters some problems in conceptualising the role of robots in human–robot interactions. In this regard, we will show how some of the concepts of ANT can help us reconceptualise both the social space in which human–robot interactions take place and their directionality. It is not our intention to apply ANT *sic et simpliciter* to the study of social robotics, but rather to gather some theoretical insights from it. The reader should therefore not expect an applicative exercise, since only a few pages of this work will remain firmly faithful to ANT.

ANT: flattening the social

Before showing how ANT can be applied to the world of social robotics, it is useful to introduce its main concepts. ANT was born around the begin-ning of the 1980s from the joint efforts of scholars working in science and technology studies. Along with Latour, Michel Callon and John Law are among the most cited scholars involved in ANT, with Latour cur-rently considered the leader of the ANT methodological project. Far from being an exhaustive, all-encompassing theory, ANT should be considered more as a set of approaches to the issue of explaining the social, sharing a number of methodological guidelines as well as some common polemical targets.

In applying ANT, we focus only on a few concepts that we consider particularly useful to our project of an anti-anthropocentric understanding of machine-hybridised social spaces. These are the often misunderstood concepts of *actor* and *network*, the fundamental problem of the "sociology of the social", and the project of "flattening the social". All these elements, we believe, can show how ANT allows us to radically rethink the social space without touching the slippery field of the ontological attribution of agency to objects. We will then depart from ANT in Chapter 3, where the limitations of ANT will be discussed in terms of the concept of "plasma".

Before going further, it is important to introduce a fundamental concept of ANT, methodologically prior to the application of ANT in the social hybridised by robots. Latour's work, in fact, extensively discusses the reasons why it is necessary to go beyond classical sociological approaches – what he calls the "sociology of the social" – in order to land in the "sociology of associations".

In his book *Reassembling the Social* (2005) which summarises and systematises Latour's many contributions to ANT, he makes an observation that is both obvious and unprecedented: a large part of sociology has confused what it had set out to explain, the "social" and its forces, with the theoretical instrument with which the explanation was to be produced. It mistook the *explanandum* for the *explanans*. The task of sociology was essentially to give an answer to what constitutes the defining "surplus" of the social: the mechanics, dynamics, and devices that make a group of individuals something more complex than the simple sum of its parts. In order to do this, however, sociology has surreptitiously assumed the "social" as a force that to some extent precedes the very existence of associations of human beings. The object that was supposed to be explained, the social, became exactly what was conceptually recruited to explain the dynamics of the world. The occurrence of a certain event was traced back to the mechanics of power, when the very target of sociology consisted of explaining the "mechanics of power". In our opinion, such a critique strikes at the heart of a certain "comfort" in which critical theory – using such concepts as ideology, devices, and social forces – has found itself. Critical theory often starts from the idea that a series of devices, such as those concepts just mentioned, can describe social events in the world, such as the occurrence of a certain social clash. However, if we look closely, we see that this strongly resembles a surreptitious operation wherein the order of the addends has been inverted: it was instead from the observation of social actors' actions that a certain mechanics was originally to be inferred. The presupposition instead of a series of "social forces" that explain events leads to some theoretical drawbacks. Firstly, the motivations of the actors become irrelevant as their accounts of their own motives are described as a case of false consciousness, granting "veritative priority" to the social theory used for the analysis. Secondly, resorting to a source of explanation – the social forces – that is presupposed with respect to reality obviates the need for proof, since it cannot be disproved by anyone – the actors themselves are a second-level source of truth. We could say that, for Latour, the sociology of the social falls into the fallacy of *petitio principii*. In fact, it explains the real with a force supposed as external, namely the social. This typically leads to theories that, by mobilising a certain conceptual core referring to supposed "social forces", are able to explain any event procedurally without adapting to different historical contexts. In short, if the facts won't fit the theory, so much the worse for the facts. As a consequence, social forces, being abstract, are hardly discoverable in local interactions. The interaction between two subjects, for example, hardly allows a focus on the social force "class struggle". If human interactions structure the social, in accordance with classical sociology, we should be able to follow the evolution and perhaps the magnitude of an explanatory device (such as class struggle) all the way up from a single interaction between two individuals to large mass dynamics. And yet, Latour notes, this is not the case: the explanatory device fails in the local dynamics. This creates a sort of irreducibility of the general to the local – of the mass dynamic to individual interaction – which reveals the confusion between the object of sociology and its instrument of explanation.

This first polemical objective of Latour's leads to a manifesto: "*flattening the social*". There is in fact no "social force" that looms like a Leviathan, no power mechanics that actually precedes the actions performed in a given context. A clear example of "flattening the social" is Latour's description of the social force called "global finance". This is no longer described as a sort of invisible reservoir of energy that, untraceable in the world of relations and objects, determines the lives of individuals. On the contrary, it is a precise set of relations between human and non-human actors, supported by transoceanic fibre optic cables, network infrastructures, computers, and "https" (hopefully not only http!) protocols for transferring data packets. Each of these actors, and many others, relates to the others in constituting the social item called "global finance" – in Latour's language, one can trace their associations. This means that a certain set of relations will be:

1 stronger than another according to the level of interconnectedness between actors (Latour, 1996) and
2 dependent on the interconnection between actors.

Quite simply, global finance ceases to exist if data exchange protocols do not function as they should – even more so if the transoceanic backbones are damaged.

What is overcome when the social is flattened is the pair of distinctions between distant/near and large scale/small scale (Latour, 1996). The first distinction loses its meaning as it is no longer relevant whether two actors are physically close since even if they are, their association can be much sparser than that of two distant actors. The degree of connection between two actors is no longer about the physical distance between the two but rather about belonging to a certain network that can be traced.

Similarly, there are no networks on a "different scale" from others, only networks that are better connected or connected with more actors. In the words of Latour (1996):

> The notion of network allows us to dissolve the micro-macro distinction that has plagued social theory from its inception. The whole metaphor of scales going from the individual to the nation state, through family, extended kin, groups, institutions etc. is replaced by a metaphor of connections. A network is never bigger than another one, it is simply longer or more intensely connected.
>
> (p. 371)

What ANT is ultimately arguing against is the supposed difference in level between abstract, non-local social forces, and the set of local relations among actors. ANT says: these two planes are the same. To separate them is to create a sort of "explicability supply" that enables the social sciences to use gnoseological devices without taking responsibility for their analytical validation. In the case of global finance, the application of such an explanatory device must therefore be justified by the traceability of associations from the Wall Street stockbroker to the social context (say, a riot) that global finance is supposed to be influencing. As we see, global finance has thus become another location on the same level as the riot. In order to connect these two locations, sociology must now trace the associations justifying its theoretical operation. To avoid the operation of splitting the global from the local: that is the sense in which ANT "flattens the social".

This does not mean that existing models for explaining social dynamics, such as classical sociology or critical theory, are unusable. In our opinion, concepts such as the Foucauldian power–knowledge device or the concept of ideology can still find a place in ANT. However, each time they are employed to explain the social we must, in Latour's words, "pay the price of the ticket" by tracing the validity of the explanation within the local interconnections of the social actors.

This approach enables new and interesting theoretical scenarios, such as the analysis of power structures, starting for instance from the differences between "http" and "https" encryption of data packets travelling on the web.

Robot mediators

From this starting point, ANT can integrate non-human actors within the social space. This integration begins with the distinction between intermediary and mediator rather than that between agent and objects. In Latour, the difference lies in the fact that an intermediary carries information without producing any kind of alteration in its meaning. A telephone records the voice of a subject in order to transmit it through radio waves, with a precise protocol that allows the signal to reach the interlocutor's device and to be reproduced as faithfully as possible – at least faithfully enough for the words to be properly understood. When delivered through a mediator, however, the message undergoes a transformation, and the output no longer corresponds to the input. A computer infected with a virus may send a large number of e-mails to one's contacts that one would never oneself have sent. In this case the computer has become a mediator: it has no longer simply transmitted information faithfully but has introduced an unexpected variable into the system. Linked to this differentiation is the concept of the agency objects take on if they become mediators of the social. It is useful here to note some implications emerging from the concepts of intermediary and mediator.

Firstly, as should be clear, the notion of agency is here completely emptied of its ontological dimension: what happens *in foro interno* of the mediating actor is irrelevant. The only relevant dimension is the performative and factual dimension of the mediation. Moreover, the fact that the agentiality of an object is dependent solely on the presence of systemic modification of a given social context gives ANT a fundamentally non-anthropocentric posture. In fact, a human actor who slavishly performs his function as a clerk by protocolling documents is "acting" less than an object that, for whatever reason, goes beyond its expected function, producing an unexpected effect. The social agency of such an object is, in principle, equal to that of a (not entirely slavish) human since it produces a systemic modification of some kind, and this conclusion is reached with no need to engage with such constructs as intentionality, free will, or other conceptual categories of the classical investigation of the social. Another aspect of agency that emerges from these premises is its eminently systemic and socio-technical nature: in this sense, there is no such thing as an action unrelated to a specific social system, and it is not possible to evaluate agency except in relation to a system. An agent is an actor who dislocates or modifies the equilibrium of a given social system, breaking the previously existing networks of relations and the socialised narratives about their reciprocal relations.

The notions of mediator and intermediary, as mentioned, are not in any way intended to solve the ontological problem of agency but only to reorder its phenomenological–social aspect. ANT is far from any kind of anthropomorphism or animism, like transferring agency onto objects in a purely "metaphorical" form borrowed from human action (Volonté, 2017).

Obviously, it follows that the more ways an actor has of modifying a given social system, the more likely it is to become a mediator. The example of social robots is in this sense of great interest. Indeed, it is difficult to think of an object/actor that is more likely to bring about unexpected changes within a socio-technical system than a talking machine. To better understand the uniqueness of the social robot within the panorama of objects, it is useful to use a comparison in line with Don Ihde's distinction between the three modes of relation between subjects and objects (Ihde, 1990). In *embodied* relations, technology becomes part of our perceived body (like glasses) when we interact with the outside world. In *hermeneutic* relations, artefacts allow us to access information about the world that is otherwise unavailable, as with the microscope. Finally, we find the *quasi-other* relationship, where the object appears to the human being in a dimension of otherness, escaping strict objectuality:

> What the "quasi otherness" of alterity relations does show is that humans may relate positively or presententially to technologies [...]. Technologies emerge as focal entities that may receive the multiple attentions humans give the different forms of the other. In alterity relations there may be, but need not be, a relation through the technology to the world [...]. The world, in this case, may remain context and background, and the technology may emerge as the foreground and focal quasi-other with which I momentarily engage.
>
> (Ihde, 1990, p. 107)

Like Ihde, Latour also notes the emergence of a phenomenological dimension of quasi-alterity in the mediating object:

> In the new definition it's just the opposite: human members and social context have been put into the background; what gets highlighted now are all the mediators whose proliferation generates, among many other entities, what could be called quasi-objects and quasi-subjects.
>
> (Latour, 2005, p. 238)

This quasi-subjective nature belongs to the human being's perception that the object has emerged from the dimension of intermediary, moving from the background to the proscenium of the social. By its very nature as an interacting object, the social robot moves out of the background of sociality to take on a full quasi-other function in social interaction. In order to justify this claim, it is useful to analyse the peculiar characteristics of the social robot, which structurally distinguish it from the technological objects we generally interact with. Indeed, we believe that the scale of quantitative differences that social robots, and future interactive technologies, present in social interactions may produce a qualitative gap, which places social robots at the beginning of a new era of (not yet fully developed) socio-technical systems.

Ihde's distinction between the various phenomenological categories of relationships with technological objects (Ihde, 1990) is useful for structurally distinguishing the interactional posture of social robots from that of other technological objects that already impact our socio-technical systems. Most of these, in fact, are supposed to be in the "background": the functions that make them "intermediaries", or *media*, are along the lines of enabling, fluidifying, or facilitating. For example, social networks help users find others with whom they share some affinity, as Facebook was created as a tool to find old school or university friends who had moved who knows where. The subject therefore does not interact *with* the technological object but instead interacts *through* it. This leads to the consideration that the object, in this case the social network, becomes a mediator only in the event of a malfunction or a sudden, unexpected interruption. As anyone who smokes knows, few objects create sociality as well as cigarettes: they can only be smoked outdoors (sigh), and their consumption has a precise duration. Generally, the place where they are consumed, far from participants in the social group who are annoyed by the smell, allows small groups of (smoking) people to interact in an unprecedented way. This sort of interaction would probably not arise if a good reason, the cigarette, had not created a new social situation. Most technological objects echo the cigarette in its function as an enabler and fluidifier of interactions. In cases of interruption, or unexpected error, the object can either stop the interaction, as in the case of a chat that ends for an application crash, or enable a new one, like a lighter that runs out of gas and forces the smoker to ask for one in the street. It is important to underline that both activities, ceasing and enabling, are mediating activities as they bring about a substantial and unexpected modification to the previously existing social organisation.

Another class of very modern objects can interact directly – which in most cases means *verbally* – with human beings. These include, in particular, such devices as voice assistants and chatbots. In this case, it is not the object that remains in the background of the interaction but its social function: in all these objects, the goal of the interaction is elsewhere, in a function – generally assistance – which is not in itself concerned with interacting. If the social function remains in the background, the social characteristics of the object, such as the range of its possible verbal interactions, will be reduced to the principle of functional efficiency to which the interaction must respond – for example, providing the information requested by the user. As we have also observed experimentally (Antonioni et al., 2021), the greater interactivity of the object does not in fact correspond to a greater capacity to be useful for a task other than the interaction itself. The social robot overcomes both of these limitations because it presents itself as a mediator of the social. Indeed, social robots interact directly, in the form of quasi-alterity, as active, interested and learning interlocutors – although this currently applies only to a few highly developed social robots. Moreover, their social function is not set against the backdrop of another objective; rather, it essentially characterises their role.

We will return to the analysis of the robot as *quasi-other* when we discuss the intersubjective implications of this posture in Chapter 6.

In the context of Latour's programme of multiplying the mediators of the social, the social robot is a good example of an object that is difficult to tame to the function of intermediary. This multiplication of mediators allows us to withdraw from the "forces" of the social their infinite supply of agency. Such forces can now be returned where they belong, namely among objects. This is Latour's fundamental anti-anthropocentric lesson. In the next section, we will attempt to account for all the possible robot-enacted modes of mediation.

An assumption that seems to be implicit in a significant part of the HRI literature is that the only relevant interactional elements are those that successfully reproduce human-like interactions (Duffy, 2003; Kiesler et al., 2008; Luria et al., 2019; Yuan & Dennis, 2019). Non-anthropomorphic non-verbal cues and behaviours are considered only to limit and reduce social interaction. For example, in current *embodied social agents*, the lack of facial expressions and body gesture coordination consistent with verbal communication is considered only as a diminution of the robot's interactivity, precluding the artefacts from socially meaningful interaction (Tinwell & Sloan, 2014). Consequently, the elements differentiating the robot's interactional capabilities from human ones become elements of the robot's non-interactivity (Bartneck et al., 2020). Therefore, they are considered only to constitute a decrease in sociality.

We believe that this approach offers a limited view of the implications of human–robot communication and that, on the contrary, for a complete understanding of HRI, those elements that deviate from or fail to faithfully reproduce a human-like interaction are equally relevant in modelling the interaction. They produce peculiar effects in the robot's posture in that specific interaction, which are by no means limited to making the relationship quantitatively less interactive. Non-anthropomorphic ways of interacting create new interactional settings that are qualitatively different from those existing in human relationships. In short, every element of an interaction, including its failure or inability to mimic an anthropomorphic behaviour, is a significant part of the interaction itself. This thesis has already been put forward in another work (Bisconti, 2021). We argued this claim through the interactionist and systemic psychology of the Palo Alto school, of which Paul Watzlawick is a well-known exponent (Watzlawick et al., 2011). To briefly review the central thesis of the paper, we argued that the systemic claim "it is impossible not to communicate" explained very clearly how a communicative incoherence on the part of the robot would, under certain conditions of anthropomorphism, be perceived as equally significant on the metacommunicative level by the human subject. For example, in order to communicate happiness on meeting someone, one must perform three communicative acts in a certain order and with a certain timing: waving, smiling, and saying "nice to meet you". The absence of one of these acts in the communication does not produce a reduced perception of joy but something completely different. A person who says "how nice to see you" in a flat, monotone voice is probably thinking the exact opposite of what he or she is saying: the use of conflicting communicative registers is a typical communicative strategy to deny the verbal content (Palazzoli & Boscolo, 1994) or to convey an ironic intention (Mizzau, 1986; Raskin, 1979). Despite the robot's inability to manage the metacommunicative side of the interaction, the human being still interacts on the metacommunicative level, where communicative acts (verbal and non-verbal) only make sense on a systemic level. Every non-verbal signal conveys a precise metacommunicative content that acquires meaning only in relation to the other elements of the interactional system (Palazzoli et al., 1993). For example, eye contact is widely regarded as one of the fundamental elements in a robot's establishing meaningful social interactions (Balistreri et al., 2011; Nomura & Kanda, 2015). However, eye contact is also considered a factor that causes relational anxiety (Schneier et al., 2011). As humans, we can tell if a subject dislikes eye contact through a series of signals: the interlocutor looks away, moves to the side, stops communicating, etc.

We take this argument a step further by arguing that consistency between different communicative channels is a crucial element for effective interactions. Therefore, when analysing the effectiveness of a specific interactional behaviour such as eye contact, we should be aware that in some cases it increases the effectiveness of the interaction. For example, eye contact when listening to a friend's emotions acquires the meaning of emotional closeness. But we should also be aware that in some cases it can cause fear and anxiety – for example, when a stranger stares without speaking. Moreover, the absence of an expected behaviour, such as smiling when hugging a friend, is also significant and changes the meaning of all the other elements of the interaction.

This means that when the robot fails – in this case, when it is unable to manage the plan of metacommunication – it leaves the tracks of inter-mediation to take on the role of mediator. The modification in the level of interaction with the human subject is unexpected, not calculated at the level of design. Indeed, in the field of HRI there are beginning to be doubts about the very validity of human–robot interaction experiments, as the results seem continually to be different, mutually incomparable, and dif-ficult to replicate. This "centrifugal" quality that the social robot brings can simply be traced back to the strong mediation that the robot enacts in the relationship. Certainly, in the current discussion (Mutlu et al., 2009), it is argued that humans attribute intentionality to non-verbal cues such as eye movements. Other scholars (Dautenhahn et al., 2006; Mumm & Mutlu, 2011; Walters et al., 2005, 2007) emphasise the importance of personal space invasion in the effectiveness of HRI; Dag Sverre Syrdal and colleagues (Syrdal et al., 2006) discuss user comfort in relation to different directions of approach by a robot. The current literature seems to consider each element of interaction – personal space, gaze, direction of approach, speed of movement, facial expressions – as separate from the others. It seems to take for granted that, in order to achieve an anthropomorphic relationality, one must simply put together all the necessary elements: eye contact, lateral approach (Dautenhahn et al., 2006), non-violation of personal space (Syrdal et al., 2007), facial expressions, voice intonation (Briggs & Scheutz, 2016), pauses in speech (Brinck & Balkenius, 2020), and so on.

As long as the systemic nature of interactional elements is not considered in the literature, the question of the robot's anthropo-morphic nature will be relegated to the juxtaposition of elements that are anthropomorphic in themselves. Yet, these elements will systemically produce relational structures different from the communicative dynamics of human relations. Nonetheless, it is reductive to think that the inter-action dynamics put forth by the robot can be discredited exclusively for being distant from an anthropomorphic interaction and leading to an unsuccessful, almost monkish interaction. Certainly, in some cases the interaction ceases because the robot deviates from fully anthropomorphic interactional settings (Satake et al., 2008). However, in our opinion it is more interesting to investigate what kinds of new relationships can emerge on the basis of this asymmetry in communication. For we have argued that juxtaposing verbal and non-verbal communicative elements is by no means sufficient to make the relationship anthropomorphic, as these elements only make sense on a systemic level. Therefore, these interactions must be conceptualised differently. A problem then arises: if the systemic equilibrium of the robot's communicative elements is not exactly anthropomorphic, how can it be defined and conceptualised? Coeckelbergh's (2011) question reported in the introduction of this manuscript is now specified one step further. The original question was: How can we know, and how should we evaluate, the human–robot relationship? We can now say, with Latour, that in failing the design objectives, i.e. the faithful mimicking of an anthropomorphic relationship, the robot becomes a mediator: it modifies a social system in an unexpected way. Nonetheless, this point of view is still limited and partial: it assumes that the robot, albeit this time a mediator, is simply "hosted" within the human anthropomorphic inter-relational system. There is a way that things are supposed to go, decided by the human actors, and a way that they have actually gone, due to a performative imperfection of the robot. Such an approach still creates the social from the normative anthropomorphic standpoint of relationality: what goes beyond it is negatively or positively contaminating the anthropocentric relational structure, which ontogenetically precedes the presence of objects in the social system. Once again, this surreptitious operation of considering the anthropomorphic relation as the standard and the starting point, which can at most be deconstructed, in our opinion does not affect the anthropocentric position at all. In fact, if we start from anthropomorphic relations, we will necessarily see the process of de-anthropomorphisation as teleological, starting from an original anthropocentrism and then gradually deconstructing it. We do not believe that this is the nature of the social systems, nor is it the aim of this manuscript: social spaces have always been inhabited by non-human actors, which have always acted. This action has, however, been concealed and covered up, or its causes misattributed, in order to maintain the subject–object difference and prevent anguish. In this regard, in the next chapter we discuss the notion of "sociomorphic interaction", which, we believe, allows us not to presuppose a-historical and ontologised anthropomorphic relations.

References

Antonioni, E., Bisconti, P., Massa, N., Nardi, D., & Suriani, V. (2021). Questioning items' link in users' perception of a training robot for elders. In H. Li, S. S. Ge, Y. Wu, A. Wykowska, H. He, X. Liu, D. Li, & J. Perez-Osorio (Eds.), *Social robotics. ICSR 2021. Lecture notes in computer science*, vol 13086 (pp. 509–518). Springer. https://doi.org/10.1007/978-3-030-90525-5_44

Balistreri, G., Nishio, S., Sorbello, R., Chella, A., & Ishiguro, H. (2011). Natural human robot meta-communication through the integration of android's sensors with environment embedded sensors. *Frontiers in Artificial Intelligence and Applications*, *233*(November), 26–37. https://doi.org/10.3233/978-1-60750-959-2-26

Bartneck, C., Belpaeme, T., Eyssel, F., Kanda, T., Keijsers, M., & Šabanović, S. (2020). Nonverbal interaction. In *Human-robot interaction* (pp. 81–97). https://doi.org/10.1017/9781108676649.006

Bisconti, P. (2021). How robots' unintentional metacommunication affects human–robot interactions. A systemic approach. *Minds and Machines*, *31*(4), 487–504. https://doi.org/10.1007/s11023-021-09584-5

Boysen, B. (2018). The embarrassment of being human. *Orbis Litterarum*, *73*(3), 225–242. https://doi.org/10.1111/oli.12174

Briggs, G., & Scheutz, M. (2016). The pragmatic social robot: Toward socially-sensitive utterance generation in human-robot interactions. In *2016 AAAI Fall Symposium – Technical Report*, FS-16-01, 12–15.

Brinck, I., & Balkenius, C. (2020). Mutual recognition in human-r interaction: A deflationary account. *Philosophy and Technology*, *33*(1), 53–70. https://doi.org/10.1007/s13347-018-0339-x

Coeckelbergh, M. (2011). You, robot: On the linguistic construction of artificial others. *AI and Society*, *26*(1), 61–69. https://doi.org/10.1007/s00146-010-0289-z

Dautenhahn, K., Walters, M., Woods, S., Koay, K. L., Nehaniv, C. L., Sisbot, A., Alami, R., & Siméon, T. (2006). How may I serve you? A robot companion approaching a seated person in a helping context. *Proceeding of the 1st ACM SIGCHI/SIGART Conference on Human-Robot Interaction – HRI '06, April 2005*, 172. https://doi.org/10.1145/1121241.1121272

Duffy, B. R. (2003). Anthropomorphism and the social robot. *Robotics and Autonomous Systems*, *42*(3–4), 177–190. https://doi.org/https://doi.org/10.1016/S0921-8890(02)00374-3

Feuerbach, L. (1972). *Essenza della religione* (C. Ascheri & C. Cesa, Eds.). Einaudi.

Freud, S. (1955). Beyond the pleasure principle. In *The standard edition of the complete psychological works of Sigmund Freud, volume XVIII (1920–1922): Beyond the pleasure principle, group psychology and other works* (pp. 1–64). Hogarth Press.

Gevorkyan, S., & Segovia, C. A. (2020). Post-heideggerian drifts: From object-oriented-ontology worldlessness to post-Nihilist worldings. *Das Questões. Filosofia Tradução Arte*, *9*(1), 3–18.

Harman, G. (2015). Object-oriented ontology. In *The Palgrave handbook of posthumanism in film and television* (pp. 401–409). Palgrave Macmillan UK. https://doi.org/10.1057/9781137430328_40

Harman, G. (2018). *Object-oriented ontology: A new theory of everything*. Penguin UK.

Heerink, M., Kröse, B., Evers, V., & Wielinga, B. (2010). Assessing acceptance of assistive social agent technology by older adults: The almere model. *International Journal of Social Robotics*, *2*(4), 361–375. https://doi.org/10.1007/s12369-010-0068-5

Henry, B. (2013). Dal Golem ai cyborgs. *Cosmopolis*, *2*, 1–20.

Husserl, E. (1968). *Richerche Logiche* (G. Piana, Ed.). Il Saggiatore.

Ihde, D. (1990). *Technology and the lifeworld: From garden to earth.* Indiana University Press.

Kiesler, S., Powers, A., Fussell, S. R., & Torrey, C. (2008). Anthropomorphic interactions with a robot and robot–like agent. *Social Cognition*, *26*(2), 169–181. https://doi.org/10.1521/soco.2008.26.2.169

Latour, B. (1996). On actor-network theory: A few clarifications. *Soziale Welt*, *47*, 369–381.

Latour, B. (2005). *Reassembling the social: An introduction to actor-network-theory.* Oxford University Press.

Luria, M., Reig, S., Tan, X. Z., Steinfeld, A., Forlizzi, J., & Zimmerman, J. (2019). Re-embodiment and co-embodiment. *Proceedings of the 2019 on Designing Interactive Systems Conference*, 633–644. https://doi.org/10.1145/3322 276.3322340

Mizzau, M. (1986). *L'ironia: la contraddizione consentita* (Vol. 29). Feltrinelli.

Morton, T. (2011). Here comes everything: The promise of object-oriented ontology. *Qui Parle*, *19*(2), 163. https://doi.org/10.5250/quiparle.19.2.0163

Mumm, J., & Mutlu, B. (2011). Human-robot proxemics: Physical and psychological distancing in human-robot interaction. *HRI 2011 – Proceedings of the 6th ACM/IEEE International Conference on Human-Robot Interaction*, 331–338. https://doi.org/10.1145/1957656.1957786

Mutlu, B., Yamaoka, F., Kanda, T., Ishiguro, H., & Hagita, N. (2009). Nonverbal leakage in robots: Communication of intentions through seemingly unintentional behavior. *Proceedings of the 4th ACM/IEEE International Conference on Human Robot Interaction – HRI '09*, *2*(1), 69. https://doi.org/10.1145/1514 095.1514110

Nomura, T., & Kanda, T. (2015). Influences of evaluation and gaze from a robot and humans' fear of negative evaluation on their preferences of the robot. *International Journal of Social Robotics*, *7*(2), 155–164. https://doi.org/10.1007/s12369-014-0270-y

Palazzoli, M. S., & Boscolo, L. (1994). *Paradox and counterparadox: A new model in the therapy of the family in schizophrenic transaction.* Jason Aronson, Incorporated.

Palazzoli, M. S., Cirillo, S., Selvini, M., & Sorrentino, A. M. (1993). *I giochi psicotici nella famiglia.* Cortina Raffaello.

Raskin, V. (1979). Semantic mechanisms of humor. *Annual Meeting of the Berkeley Linguistics Society*, *5*, 325. https://doi.org/10.3765/bls.v5i0.2164

Rosenthal-von der Pütten, A. M., Schulte, F. P., Eimler, S. C., Sobieraj, S., Hoffmann, L., Maderwald, S., Brand, M., & Krämer, N. C. (2014). Investigations on empathy towards humans and robots using fMRI. *Computers in Human Behavior*, *33*, 201–212. https://doi.org/10.1016/j.chb.2014.01.004

Satake, S., Kanda, T., Glas, D. F., Imai, M., Ishiguro, H., & Hagita, N. (2008). How to approach humans?-Strategies for social robots to initiate interaction. *Proceedings of the 4th ACM/IEEE International Conference on Human-Robot Interaction, HRI'09*, 109–116. https://doi.org/10.1145/1514095.1514117

Schneier, F. R., Rodebaugh, T. L., Blanco, C., Lewin, H., & Liebowitz, M. R. (2011). Fear and avoidance of eye contact in social anxiety disorder. *Comprehensive Psychiatry*, *52*(1), 81–87. https://doi.org/10.1016/j.comppsych.2010.04.006

Searle, J. R. (1980). Minds, brains, and programs. In *The Turing Test: Verbal behaviour as the hallmark of intelligence* (pp. 201–224). Bradford Books.

Syrdal, D. S., Dautenhahn, K., Woods, S., Walters, M. L., & Kheng Lee Koay. (2006). "Doing the right thing wrong" – Personality and tolerance to uncomfortable robot approaches. *ROMAN 2006 – The 15th IEEE International Symposium on Robot and Human Interactive Communication*, 183–188. https://doi.org/10.1109/ROMAN.2006.314415

Syrdal, D. S., Lee Koay, K., Walters, M. L., & Dautenhahn, K. (2007). A personalized robot companion? – The role of individual differences on spatial preferences in HRI scenarios. *RO-MAN 2007 – The 16th IEEE International Symposium on Robot and Human Interactive Communication*, 1143–1148. https://doi.org/10.1109/ROMAN.2007.4415252

Tinwell, A., & Sloan, R. J. S. (2014). Children's perception of uncanny human-like virtual characters. *Computers in Human Behavior*, *36*, 286–296. https://doi.org/10.1016/j.chb.2014.03.073

Volonté, P. (2017). Il contributo dell' Actor-Network Theory alla discussione sull' agency degli oggetti. *Politica & Società*, *1*, 31–58.

Walters, M. L., Dautenhahn, K., Koay, K. L., Kaouri, C., Boekhorst, R., Nehaniv, C., Werry, I., & Lee, D. (2005). Close encounters: Spatial distances between people and a robot of mechanistic appearance. *5th IEEE-RAS International Conference on Humanoid Robots, 2005.*, *2005*, 450–455. https://doi.org/10.1109/ICHR.2005.1573608

Walters, M. L., Dautenhahn, K., Woods, S. N., & Koay, K. L. (2007). Robotic etiquette: Results from user studies involving a fetch and carry task. *HRI 2007 – Proceedings of the 2007 ACM/IEEE Conference on Human-Robot Interaction – Robot as Team Member*, 317–324. https://doi.org/10.1145/1228716.1228759

Watzlawick, P. (1971). *Pragmatica della comunicazione umana*. Astrolabio.

Watzlawick, P., Bavelas, J. B., & Jackson, D. D. (2011). *Pragmatics of human communication: A study of interactional patterns, pathologies and paradoxes*. WW Norton & Company.

Yuan, L. (Ivy), & Dennis, A. R. (2019). Acting like humans? Anthropomorphism and consumer's willingness to pay in electronic commerce. *Journal of Management Information Systems*, *36*(2), 450–477. https://doi.org/10.1080/07421222.2019.1598691

Žižek, S. (2016). Afterword: Objects, objects everywhere. In *Slavoj Žižek and Dialectical Materialism* (pp. 177–192). Springer.

3 Beyond anthropomorphic interactions

At Aarhus University, Denmark, reflection on the novel concepts of relationship that human–robot interaction (HRI) may introduce is flourishing in the work of Johanna Seibt, which represents an interesting starting point for the discussion of the concepts of asymmetrical relationships and *sociomorphing*.

The concept of asymmetrical relationships, developed by Seibt (2016, 2017), is intended to help classify non-symmetrical social interactions – interactions where participants do not have the same attributes. Seibt starts from the claim that "we should abandon the idea of a dualist distinction between social and non-social interactions; rather, we should conceive sociality as a matter of degree".

The asymmetrical relationship thus concerns the degrees of a robot's ability to effectively simulate a certain type of action (such as greeting). These degrees can be explored by building an expansion matrix to evaluate the simulation effectiveness of each relevant action performed by the robot in the act of greeting (raising its hand, moving it, pronouncing the words of greeting, etc.). The asymmetry in the relationship stems from the fact that different degrees of simulation produce different degrees of perspective-taking by human subjects with respect to the interaction taking place. This perspective-taking is a kind of theory of mind that human agents form with respect to the robot. For example, while greeting the robot, I can think either that it understands my gesture or that it does not. These alternatives are described by as

> "types of experienced sociality" [...] a technical term standing for items in a classification of complex phenomenological contents which in first approximation can be characterised as feelings of co-presence or "being-with". Feelings of co-presence directly tie in with behaviour, but they are distinctive ingredients of human mental life.
>
> (Seibt et al., 2020, p. 59)

DOI: 10.4324/9781003459798-3

The robot, then, provides the feeling of a certain degree of sociality based on how well it manages to produce an effective and human-like "type of experienced sociality". If the robot responds immediately to my greeting, I will have the feeling that my greeting "matters" to the robot. If it does not respond at all, I will simply have made a formal gesture due to the fact that the robot, if anthropomorphic, corresponds to the form of the human beings I usually greet. The concept of "sociomorphing" builds onto this approach and, in our view, represents its most interesting conceptualisation. Seibt et al. (2020) discuss some experiments conducted by the robophilosophy unit at Aarhus. They claim that what seemed to describe the attitude of human subjects towards robots was not anthropomorphism but "sociomorphing". Sociomorphic interaction is characterised by the perception of *non-human* social abilities ("perception of actual non-human social capacities"). If the interaction can be sociomorphic, this means, in Seibt's view, that anthropomorphising robots is unnecessary, for that is not the only functional form of HRI. As we already argued above, for this reason we claim that an anti-anthropocentric perspective that looks at the role of robots in society must pass first and foremost through a redefinition of the categories defining the HRI. This theoretical redefinition should ultimately shape the normative assumptions of technology design. As we argued above, non-anthropomorphic interactional elements are nonetheless meaningful to the human subject. This follows from the systemic nature of communication. Therefore, when a robot moves away from anthropomorphism, we say with Seibt that it performs in a category of the expansion matrix distant from perfect simulation. This performance is nevertheless meaningful in the interaction and for any human subject relating to the robot. The category of *sociomorphing* is therefore certainly important for understanding the asymmetry of the interaction between humans and robots.

Nevertheless, one element of Seibt's argument is problematic: the taxonomy still develops in a linear dimension from anthropomorphism to non-anthropomorphic interactions, namely from perfect simulation of human interaction to its merely functional simulation.

The taxonomy's linearity fails to do justice to Seibt's efforts to overcome an anthropocentric view of HRI. Against this backdrop, it would be interesting to test experimentally the applicability of some typical items on the HRI questionnaires, for example *competence* or *trustworthiness*, to absolutely non-anthropomorphic objects. We would probably notice that the very shape of the objects, whether pointed or round, modifies these items as in the "bouba-kiki effect" (Peiffer-Smadja & Cohen, 2019).[1] However, Seibt defines sociomorphic interactions as grounded in and conceptualised through the expansion matrix of her previous work (Seibt, 2017). This reference to a linear evaluation of interaction might limit the effectiveness of the notion of sociomorphing. However, we believe that Seibt's work fully grasps the problem of the narrowness of the category of anthropomorphic interaction. We do not have an alternative solution at this time to the problem of describing the ontological nature of HRI. Nevertheless, we believe that a concerted effort should drive the research towards a conceptualisation of inter-action as *also* but not *necessarily* anthropomorphic. We should, however, be aware that a certain anthropocentric stance persists when we concep-tualise "sociomorphic" interactions as lying on a scale that goes from "not human" to "human".

To the *sociomorphic* notion of interaction we link and propose that of a socio-centric HRI to replace the anthropocentric perspective. As we discussed above, interactional elements are systemic; because of this, we should expect to find new forms of non-human social interaction.

An intriguing exploration and conceptualisation of human–machine interactions is presented by Federica Russo in her book "Techno-scientific practices: an informational approach" (Russo, 2022). Of particular note, and forming the theoretical heart of her manuscript, is the concept of the *poiesis* of artificial agents. In Chapter 9, Russo's argument bifurcates into two primary trajectories:

Firstly, she aims to show that artificial agents qualify as genuine agents due to their ability to process information. This capability transforms them into epistemic agents, which can alter the dynamics between other agents and their environment. Russo's perspective on epistemic agency is inherently relational. Knowledge, in her view, emerges from the col-laboration of all agents capable of processing information. These agents can influence the environment, introducing new dynamics between it and other agents.

Secondly, to elucidate the partial autonomy and consequent "agentiality" of technical objects, Russo engages with the works of both Latour and Gilbert Simondon. The discussion on Simondon's work will be further explored in the subsequent chapter, which delves into the con-cept of metastability. Russo's stance on epistemic agency is anchored in constructionism, a theoretical approach that bridges the gap between epi-stemic realism (where knowledge mirrors an objective external reality) and constructivism (where reality is crafted *ex nihilo* by epistemic agents). Specifically, constructionism posits that knowledge is both relational and distributed among the epistemic agents within a socio-technical system. This implies that any epistemic certainty is always contingent upon a specific configuration of relations – both between the world and the epi-stemic agents and among the agents themselves. Therefore, evaluating the epistemic validity of a concept or claim is contingent upon the "level of abstraction" at which the analysis is conducted.

Knowledge is relational also at the level of the concepts or of the semantic artefacts that compose it. I speak of semantic artefacts to emphasise that we make the concepts that we use to make sense of the world around us. I explored the idea that human epistemic agents are makers not just because we make artefacts, but also because we make semantic artefacts or concepts. To say that knowledge is relational at the levels of concepts means that these are not islands but are always connected to other concepts. I take this to be an irreducible relational aspect of knowledge.

(Russo, 2022)

In our view, the pivotal insight of this conceptual framework is the notion that knowledge, including concepts, is an artifact birthed from the epistemic collaboration between poietic agents, both human and non-human. We resonate strongly with this perspective and intend to incorporate Russo's viewpoint in the ensuing chapters of our manuscript. We posit, with Russo, that knowledge is not merely constructed by all epistemic participants but should be regarded as an artifact circulating within the interaction system of various poietic agents. Our goal is to delve deeper into interactions perceived as a system of signifiers exchanged by actors. These signifiers encompass all semantic artifacts discerned within a social system by at least one participant. The significance of a semantic artifact arises solely from its relation to other elements within the system. Some configurations of that semantic equilibrium will be anthropomorphic, others will not. All configurations in general will be sociomorphic from the moment that the action performed is supposed to be significant for someone, human or not.

This argument does not require any ontological notion of intentionality, and it aligns with Latour's assumptions in ANT. In the next section we expand this idea, applying it to social interactions populated by robots.

Dislocating anthropomorphism

One of the main limitations of the current approach to HRI, from the point of view of the philosophy of robotics as well as others, is that it typically considers HRI in terms of its "intersubjective" structure (here in the sense simply of a one-to-one relationship). HRI scholars often forget that, except in laboratory conditions, the subject is unlikely to be interacting with a robot alone. This is true for almost any kind of robot: in social assistive robotics for the elderly, the robot – be it an anthropomorphic NAO or a zoomorphic PARO – typically interacts with elders who, in the meantime, are interacting with each other. This issue is fundamental to understanding, outside of laboratory dynamics, what intersubjective and social conditions the robot is actually facing. The second part of this manuscript will be concerned with outlining the effects and consequences of *companion robotics* on the intersubjective level, given the current anthropocentrism of normative design practices and policies.

For now, however, we continue in our present enterprise, namely the quest to understand how social systems hybridised by non-human actors are structured. While the agency of the object has been elided in order to mark a chasm between human and non-human, the social robot, in contrast, has been surreptitiously included and normed in the form of the *anthropos*, its interactions constrained in the narrow gnoseological meshes of anthropomorphism. As we discussed in the introduction, the distressing problem of differentiating oneself from the object does not entail only pushing the *ens* back into the category of object in order to ensure that social agency remains a distinctively human category. If the social robot happens to enter human sociality as a fully recognisable actor, then it must immediately be included in the anthropomorphic category. The point of anthropocentrism, we believe, is not only to secure the human being a prominent place in the cosmos of actors. Above all, its aim is to nullify everything between the two very distinct categories of subject and object.

The inclusion of a specific type of object, such as a robot or AI system, within the category of anthropomorphic agent is much less destabilising than including non-anthropomorphic forms of agency. The former case demands only sharing the space of social agency with someone else, who, however, has already been brought back to "anthropo-normed" categories. The latter case, on the other hand, breaks the rigid line of distinction, and subjects are in constant danger of being displaced. We believe that a large part of the current discussion in the field of philosophy of robotics, and in particular in the field of machine ethics, is currently making a huge attempt to adapt attributes of subjectivity to interactive social robots for this very reason. We believe that the discussions, for example, on moral machines (Awad et al., 2018) are all epiphenomena of this ideological–theoretical movement.

In the second case – the inclusion of non-anthropomorphic forms of agency – we should instead admit that the dimensions of agency are multiple and perhaps not even fully enumerable. The latter option produces the collapse of the anthropocentric structure as well as of the reassuring distinction between subject and object. Against this backdrop we agree with Barbara Henry's (2013) claim:

> It is this "power to normalize entities" that allows us humans to cognitively dispose of the liminal zone between life and non-life, between nature and artifice.
>
> (p. 3)

This power to "dispose of" that liminal zone stems from the possibility of managing and ordering an imaginary. The question of who manages the social imaginary, who produces it, is obviously eminently political-normative (Henry, 2014) and connected with what we pointed out in the introduction regarding the "war of inscriptions" for the colonisation of the robot body. But the role of narratives, as we will also discuss in the following pages, does not only concern the appropriation of the *entia* in a political-normative way. There is also a relevant symbolic register to which the sci-fi discourse concerning the "undead" is geared. This, as Henry points out, produces an imaginary where the Golem, in the archetypal form of the posthuman cyborg, is

> a liminal, or if we prefer, ambivalent place, in which – depending on the perspective of the beholder – life and death, with the change of their respective positions, affect the imaginative configuration of multiple and differentiated organised and sentient humanoid entities.
>
> (Henry, 2013, p. 2)

In this sense, we can already anticipate some of the conclusions of this work regarding the intersubjective status of human–robot relations: if the cyborg represents the hybridisation between human and non-human, the robot is instead the social mediator that "sociomorphises" previously anthropomorphic social interactions. What becomes post-anthropocentric is neither the robot nor the human being but the space of interaction. It is the social body, and not the robot's body, that enters into metaphorical redundancy with the cyborg's body.

Who is enrolling whom?

When discussing how robots become mediators, we claimed that communicative error produces an unintended mediation on the part of the robot. Further on, we found the peculiar forms of communication that the robot enacts on the level of metacommunication. We have discussed how the modification of metacommunicative balances leads the robot to convey peculiar and unusual communicative structures. All this can be included under Seibt's concept of *sociomorphing*, where anthropomorphic interactions can be considered one of several sub-classes of sociomorphic interaction. Finally, the concept of "poietic agent" clarified the co-construction of knowledge between humans and non-humans.

Therefore, we briefly depicted the mediating function of the robot in the intersubjective relationship, which is even more important and visible outside the sterile observational setting of the laboratory. Experiments with

HRI in open spaces and uncontrolled contexts clearly show the ability of robots to enrol subjects, create networks, and become spokespersons.

But what does it mean to be enrolled by a robot? The first idea that emerges, to those familiar with *interaction studies*, is certainly the dimension of engagement. The more the subject is engaged by the robot in the social interaction, the more the robot "enrols". This engagement is multi-directional along at least three lines: from actor A to actor B, from actor B to actor A and from actors A and B to actor C. The following example illustrates these three directions.

Consider an assistive robot for the elderly in a nursing home, a typical use case in human–social robot interactions (Heerink et al., 2010; Sharkey & Sharkey, 2012; Turkle et al., 2006). Assuming that the robot is a NAO, in this case the three directions are:

1 The robot engages in the interaction with the human being.
2 The human engages the robot.
3 The human engages another human via the robot.

What should be noted is that none of these interactions is purely anthropomorphic. Furthermore, in the third case, the mediation process is carried out through a non-human actor, who modifies the structure of the relationship between the two non-human subjects with sociomorphic communicative elements. This situation is not the same as that of social networks, which remain in the background as they mediate human interactions. In this case, the social robot is a constitutive actor in the social system. Sociomorphing, namely a non-anthropomorphic setting of the interactional elements in the given social system, is modifying the structure of relations between human beings. Recalling Latour's notion of "mediation", we see that the social robot to some extent modifies the association of networks, both human and non-human, sometimes acting as their spokesperson. One example is robots that relate not to an individual user but to a group of subjects, increasing the sociality of a group through their interaction (Kachouie et al., 2014). The function of mediator is amplified by the sociomorphic nature of the interaction, since an excessively anthropomorphic structure of interaction would necessarily reduce the "entropy" of the social system in which the robot is situated, making interactions largely more predictable. It is precisely because the robot "makes mistakes" – i.e. produces interactions that are not typical of anthropomorphic relationships – that it may be able to increase the system's sociality. What in a dual interaction is (perhaps) a limitation of current robotics is a strength in actor networks: the inconstancy of the robot with respect to relational anthropomorphism allows a faster and more dynamic reformulation of networks. Social ties are thus partly sustained by robots themselves, precisely because of their *quasi-other* nature (Latour, 2005). Finally, this term acquires a definitive meaning within a non-anthropocentric view of social robotics. The quasi-otherness of the robot refers not only to the robot's being midway between subject and object but also and above all to its being sociomorphic in the inter-action. It is a quasi-other because it disturbs the idea that "otherness" in a social system can only be anthropomorphic. As we have already suggested in another work (Bisconti & Carnevale, 2022), anthropic communicative structures increasingly stress the possibility of entering into a quasi-alterity relation with the object, in accordance with Ihde's (1990) proposal. This means, from a socio-centric perspective, that the *quasi-other* is one who acts as a mediator using forms of action extremely similar to human ones – the most obvious case being that it speaks our language – while neverthe-less remaining to some extent distant from anthropomorphic interaction.

Nevertheless, a doubt remains in this discussion: what makes an actor become a mediator, departing from the function of stable intermediary? According to Latour, it seems that the mediator does nothing more than act unexpectedly. The only real element that differentiates the medi-ator from the intermediary is the failure to faithfully carry the entrusted message. Yet such an explanation is not sufficient to account for the whole sphere of aggregations and disintegrations of actors in their incessant and apparently disordered flowing that ultimately reorganises a new system configuration with new spokespeople. What seems to be missing – and this is the topic of the next section – is the whole horizon of meaning that makes the connections relevant to the actors in play. Latour has done out-standing work in explaining what operations must take place in the real world in order for a given significant content to be conveyed or taken up by an actor. However, it is still unclear under what conditions a specific content changes an intermediary into a mediator. When a robot relates to human actors in a certain social context, the set of actions it carries out consequently takes on meaning for the other actors in the context. We define socio-symbolic artifacts as a series of meaningful interactional elem-ents that acquire a value on the basis of a systemic equilibrium: each one specifies the next, and the previous in turn takes on a value only in rela-tion to all the other interactional elements present in the social context. In ANT, not every act generates a modification, a reshaping of the networks that are constantly forming and unravelling. Under what conditions, then, does a robot's act create a significant modification of the social space? In the next section, we attempt to solve this problem.

Networks of meaning

Why do fierce armies disappear in a week? Why do whole empires like the Soviet one vanish in a few months? Why do companies who cover the world go bankrupt after their next quarterly report? Why do the same companies, in less than two semesters, jump from being deep in the red to showing a massive profit? Why is it that quiet citizens turn into revolutionary crowds or that grim mass rallies break down into a joyous crowd of free citizens? Why is it that some dull individual is suddenly moved into action by an obscure piece of news? Why is it that such a stale academic musician is suddenly seized by the most daring rhythms? Generals, editorialists, managers, observers, moralists often say that those sudden changes have a soft impalpable liquid quality about them. That's exactly the etymology of plasma.

(Latour, 2005, p. 245)

Towards the end of *Reassembling the Social*, Latour thematises a doubt that seizes the reader right from the opening pages of the book. This magnificent flow, materialisation, and dematerialisation of associations between actors following the rules of attachment, plug-ins, etc., does not seem to answer one question: why did this particular association of actors form, and why did that other one dissolve? The principle according to which networks form prevails over other networks and are finally replaced with still other configurations remains unknown. This is a central problem because it makes actor–network theory (ANT) only half a descriptive theory: it can describe how a phenomenon occurs, but without explaining the principle that causes the phenomenon. Latour offers a lens of outstanding clarity for considering the social as always hybridised by the non-human, by the objectual (Latour, 2002).[2] However, we are not yet able to explain the process that leads a network to acquire a specific configuration with a specific spokesperson. Nor are we able to explain why that given, specific association of actors unravels and is replaced by another. Moreover, in the excerpt just quoted, Latour presents another question: why do structural modifications of the social take place in a very few moments, hours, or days, after extended periods of apparent quiet? Why does the arrival of new mediators, who apparently bring about a minimal change in the configuration of the social system, sometimes produce the collapse of entire social systems that were stable until the day before? The shift from the topic of the actor to the topic of the network moves the focus from the problem of mediation to that of associations (Volonté, 2017a,b); however, it still remains unclear on both levels what gives rise to the crisis. The concept of the agency of actors is completely reinterpreted by Latour at the systemic level. Here "agency" is related not to actors' intentionality but to the modification it brings to the web of associations. Still, the problem of how one becomes a mediator – and therefore also of how a network is formed or unravelled – remains unaddressed. At the level of the mediator, this concerns the characteristics that the information carried by an actor must have in order to mediate the social, reassembling it in a new configuration. It would in fact be simplistic to claim that information can be transported by an actor either faithfully, as by an intermediary, or unfaithfully, thus becoming mediated. If the possibility of carrying information without modification ever existed, which in our view is hardly tenable, this would in any case happen in a negligible number of cases. Every participant in the interaction, to some extent, brings about a systemic change of some kind.[3] So what needs to be investigated is the nature of the change to the information that leads an actor to function as a mediator and a network to be created or unravelled. The answers to these questions are absent in Latour, but the issue is formalised in the concept of "plasma".

> It seems that no understanding of the social can be provided if you don't turn your attention to another range of unformatted phenomena. It's as if at some point you had to leave the solid land and go to sea. I call this background plasma, namely that which is not yet formatted, not yet measured, not yet socialised, not yet engaged in metrological chains, and not yet covered, surveyed, mobilised, or subjectified. How big is it? [...] "it's astronomically massive in size and range".
>
> (p. 244)

Plasma is thus like a liquid – Latour compares it shortly afterwards to the sea – that fills the conduits of associations between actors. The network provides a precise image of what Latour seeks to express with the concept of plasma: the distance between the threads of the network, however small, constitutes a certain void. This void, which indicates the theoretically unbridgeable distance between one actor and another, provides at least the background, an intuition, of the problem. In Chapter 4 we conceptualise the nature and role of the plasma. We will take Latour's work further by attempting to answer the question of the nature and functioning of crises in social systems.

Meaning-making processes, panoramas, attachments, and plug-ins

Within ANT, the question of language seems completely abandoned and an irrelevant element in the constitution of the social, the actors and their associations. Volonté (2017a,b) already notices this absence:

> I think, however, that part of this agency is modulated through the process of meaning-making [...]. It is as if the hybrid character of each actor-network does not also include the meaning that humans attribute to their field of experience, as if language were transparent to its meanings, as if the accounts of the protagonists of a situation were

merely descriptive. It seems to me, therefore, that here the space opens up for an extension of the investigation into the agency of objects in the human world, in which to study the relationship between the processes of constituting meaning and the exercise of agency by material objects.

(p. 54 [Translation is mine])

We gladly accept the invitation to extend the investigation and agree with Volonté, who underlines how the question of meaning-making is completely ignored in ANT. A few pages above, we pointed out how it is hardly sufficient to ground the principle of mediation in the mere unexpected action of one of the actors, who instead of conveying the message faithfully alters it to some extent. That solution has two limitations. The first is to reduce the question of mediation in the social sphere to the mere dimension of error, thus rendering it fundamentally a stochastic process or little more. This solution seems unlikely because it would entail that mediation processes follow a completely uncoordinated, and in a certain sense "anti-social", flow. How could networks of associations be created solely on the basis of errors in the transmission of the message? On this view, any process of mediation, and thus of alteration, would have only a certain (minimal) chance of creating, rather than unravelling, a network. The entropy of mediation would therefore make the dissolution of networks more likely in the extreme than their formation. This brings us to the second limitation: the number of occasions on which information is carried in a completely faithful way is lower to a staggering degree than the number of those in which it undergoes some modification, even if minimal. This is also argued by Latour: the multiplication of mediators, the primary goal of ANT, consists precisely in following the actors closely enough to notice how very frequent the process of mediation is. If the mediation process is entropic and alterative, the only way in which a network remains stable is through intermediation, yet intermediation is highly unlikely. This poses the first linguistic problem of ANT, which we can summarise in the following consideration:

1 ANT offers no explanation of the processes of sense-making in relation to the processes of network constitution. The process that leads an actor to alter information is not investigated, despite being the basis of the mediation principle. What kind of alteration is therefore a mediating one?

A nearly linguistic element introduced by Latour is the plug-in, which, like a data package downloaded from the web to "enable" a computer to display a certain content, is an "immaterial object" that actors can "download" and install in the structure of their actions.

This tone of voice, this unusual expression, this gesture of the hand, this gait, this posture, aren't these traceable as well? And then there is the question of your inner feelings. Have they not been given to you? Doesn't reading novels help you to know how to love? How would you know which group you pertain to without ceaselessly downloading some of the cultural clichés that all the others are bombarding you with?

(Latour, Reassembling the Social, p. 209)

It appears that Latour's term "plug-in" aligns closely with Russo's concept of a "semantic artifact" as described in her 2022 work. While Russo delves into the epistemic implications of this concept, Latour is more captivated by its sociological essence. Drawing inspiration from both Russo and Latour, we introduce the term "socio-symbolic artifacts". This term is chosen to underscore several facets: firstly, these artifacts are collaboratively shaped by both human and non-human actors. Secondly, they extend beyond mere linguistic expressions, such as utterances, and can also manifest in non-verbal forms, akin to the clichés Latour references. As a result, these artifacts sometimes deviate from the traditional understanding of knowledge, rendering them distinct from conventional epistemic artifacts. Socio-symbolic artifacts, in contrast, represent content that holds collective significance within a network of actors. Latour gives some examples in the quoted passage, which we can extend to religious beliefs, moral values, political convictions, etc. Each actor would accordingly "download" a series of informative contents, which would enable recognition within a social network. A group of actors sharing the same plug-ins would produce a *panorama*, another recurrent term, designating what we might call the self-narratives of networks. With a function similar to the concept of "ideology", Latour's panoramas are horizons of meaning produced by networks, often competing with each other in an attempt to prevail in describing the relationship between networks and the real that exists outside.[4] Panoramas thus exist for the dual purpose of describing the real and becoming the semantic–normative structures that allow a specific network of associations to enrol new actors.

As we can see, the topic of language is not totally absent in Latour, but it is constantly kept in the background and never addressed thematically. If actors "download" plug-ins, these must first of all be something ontological. Are they also objects endowed with "agency"? Are they therefore also actors, capable of mediation? While before the social was perfectly flattened, these second-level objects, basically structures of signification above the real, partly upset the universe of ANT. But, beyond this, the main issue is the very process of acquiring these symbolic structures. Even assuming, with Latour, that actors are indeed in a position to "freely" acquire a specific plug-in, e.g. adherence to the rituals of Catholic ceremonies and the consequent definition of oneself as a Christian, what process guides the choice of one such plug-in over another? This is not explained by ANT, which, to account for actors' behaviour, only mobilises the mediation principle, invoking information altered by another actor. Is the modification made by the actors in their own interest? Is it in the interests of the network of associations to which they belong? And, above all, what does "interest" mean in ANT?

The same problem concerns the notion of panoramas, which are narratives constructed by a specific network to describe the entire set of existing "networks". Panoramas can thus be described as plug-ins within a network of associations that are so extremely viral that they are "downloaded" by most of the actors in the network.[5] Hence the second consideration:

2 ANT does not describe the principles of circulation of socio-symbolic structures within a network, nor does it explain the mechanisms leading to the enrolment of actors.

These two considerations lead us to wonder what the actual process is that constitutes the actor–mediator.

To sum up, we are raising the same issue from two perspectives. The first is that of the mediator, who performs one specific action (acquiring a plug-in, etc.) instead of another: for what reasons, and through what process, is a specific action chosen, be it mediation or intermediation? The second perspective concerns how a network can be either unravelled or formed by an act of mediation: what processes and mechanisms regulate the functioning of networks – or what alteration of the content of information constitutes a mediation, and why? What we are going to investigate, in the case of this second perspective, is what lies behind the social and its associations, understood in the original Latin sense of "socius", translatable as "companion, ally". So, what is shared in *socii* networks? Our aim is not to abandon the cornerstones of ANT in the attempt to solve these problems but to follow the path indicated by Volonté (2017a,b): that is, to explore the dimension of meaning-making processes from an actor–network perspective.

What do we need a theory of crisis for? Badiou and the dichotomy Being–Event

With the notion of "plasma", Latour identifies a core challenge for his theory. As already mentioned, Volonté (2017a,b) suggests overcoming the problem by recovering the linguistic and symbolic dimension, with the linguistic dimension playing a role in the inner workings of the mediation process and in the mechanisms of network association. For such a step, it is useful to think of the concept *crisis* ("Crisi", 1997) from the Greek κρίσις, which has the double meaning of *dispute* and *decision*. Both these dimensions are present at the moment when an actor, whether human or non-human, breaks the quiet flow of intermediations. At that moment, that actor makes a *decision* that sets up a *dispute* between the old order of relations between actors and the new order, which is formed by the mediation event. Mediation and crisis are thus two contiguous aspects of social systems, one being the cause, the other the effect.

This problem of the nature of the mediating event is eminently political. What we emphasise is the need to understand the mechanisms according to which the real is ordered and the socio-symbolic structures crystallise and circulate between actors – the reasons why great historical events, such as crises of social systems, take place at one particular moment instead of another. A theory of crisis, while appearing to be an eminently gnoseological problem, allows us to conceptualise the conditions of possibility that allow an event, such as the introduction of interactive non-human actors like social robots, to bring about an essential change within the social structure. To accomplish this, it is necessary to understand under what conditions and through what mechanisms social systems enter into crisis. Given such a foundation, we can investigate what kind(s) of crisis, and thus what kinds of mediations, non-human interacting actors can bring into social systems. For this reason, the next pages of this manuscript will focus on the gnoseological issues in the theory of crisis and its definition. Following this journey, we will return to the political issues of machine design.

In fact, establishing a theory of the modification processes of socio-symbolic structures will allow the conceptualisation of the three issues that social robotics poses at the political level: the robot as the field of the war of inscriptions/descriptions in the context of anthropocentrism; the robot as the eschatological support; and the robot as the phantasmatic support of the patriarchy's desire structure.

In *Being and the Event*, Badiou (2007) has already thematised the problem of crisis. In spite of the drastic difference from Latour in linguistic register, we can compare the action of mediation to what Badiou (p. 181) calls the *evenemential*, or more simply the Event, which gives historical structure to the frame of the happening. Outside of this, the rest is Nature, where the event is not occurring and there is instead intrinsic stability. From a political point of view, the Event designates the excess, the void at the centre, that emerges from the order of Being and to some extent identifies the set of possible experiences in a given historical order. The gap between Being and Event becomes, from a political point of view, the space where a certain "truth" of the Being appears in the *evenemential* form. This Truth is always structured as a certain lack within the set of the thinkable and symbolisable. In Badiou, the event of the French Revolution is the manifestation of the Truth of Being of the *ancien régime*, and the coming to light of that Truth is the manifestation of the emptiness at its centre. As Zizek (1999) captures in his book *The Ticklish Subject*, Badiou's Truth/Event closely resembles the concept of "symptom" in psychoanalysis, where the truth of the subject – in this metaphor, the subject's disorder – is only visible from a point outside the set of voluntary actions.

The crisis in Badiou thus occurs when a series of subjects are "enrolled"[6] by the emergence of Truth and become its spokespersons. This Being, as well as this Truth/Event, is always historicised in Badiou. The Truth is always that of a precise moment and historical context. In this sense, one could say that Badiou's is a historicised metaphysics.

Two issues emerge from this vision of Badiou, which further compli-cate the questions regarding a possible theory of crisis that we raised with Latour. First, in Badiou the problem of mediation, which takes the name of Truth (while Being would represent the chain of intermediations), leads to the total exclusion of the actors' agency. The actors are simply enrolled by the Truth/Event, which manifests itself in an apparently autonomous fashion. Badiou's theory moves our reflection forward by conceptualising mediation as a rupture, a ripple, in the order of Being/the thinkable. However, it takes us a step backwards compared to Latour, by falling into the trap of the sociology of the social that elides the role of actors from historical happening.

A second issue that quickly arises concerns the relationship between Being and Event in the intimate structure of the manifestation of Truth:

> It is the Event which belongs to conceptual construction, in the double sense that it can only be thought by anticipating its abstract form, and it can only be revealed in the retroaction of an interventional practice which is itself entirely thought through.
>
> (Badiou, 2007, p. 178)

The Truth/Event of Being therefore happens (in Badiou's quotation: "is revealed") only retroactively; it is a narration that the historical actors perform retrospectively.

As Zizek explains (Žižek, 1999):

> A neutral historicist gaze will never recognise in the French Revolution a series of traces of the Event called the French Revolution, but only a multitude of events caught in the web of social determinations.
>
> (p. 122 [Translation mine])

In fact, in Badiou an Event that disrupts the plane of Being[7] is possible only on condition of an *interpreting intervention*. This is because the interpreting intervention is nothing other than the construction of a structure of meanings jointly ascribed by the actors to a series of significant events, such as a riot or the murder of a king. These narratives retrospectively compose the symbolic–enunciative structures that come to describe the Truth of the Event. Therefore, we can make a logical distinction between Event and Truth in Badiou. The Event is a happening on the real ontological level, a reconfiguration of the material conditions of a certain observable context. The Truth of the Event, on the other hand, represents the retrospective interpretive intervention of the actors, who produce shared statements on the Event that has already happened (but that has socially happened only through this interpretative intervention). These statements are responsible for deciding whether the event that occurred is on the plane of Being (intermediation) or on the plane of Event (mediation). This means, to move forward, that the Truth/Event qualifies as such only when an element coming from outside the current narratives disrupts the socio-symbolic structure[8] of a network to some extent. In fact, if no disruption happens, the Event does not retrospectively become a Truth and does not enrol actors. In that case, the event remains in the realm of Being (intermediation) and does not become a Truth (mediation), since the actors have not retrospectively designated the event as a Truth. With these distinctions, we have advanced considerably in our understanding of the structure of the mediating Event, but we are still unaware of the mechanics that guide actors to recognise one event as Truth rather than another. In the next pages, we discuss the systemic theories of Maturana and Varela. We argue that the reason why a certain Event is chosen as the one carrying the Truth of a certain state of Being is related to the socio-symbolic systemic equilibrium of a given social system.

Starting from these considerations, in the next chapter we will discuss what mechanisms drive the reformulation processes of socio-symbolic structures.

Notes

1 As for example in the case of the two images to be associated with two words "bouba" and "kiki": the former is generally associated with a round image and the latter with an angular one.
2 This consideration raises some reflections of a philosophical and political nature on the margins of this discussion, regarding the theories and approaches that describe the world today. We believe that the operation of surreptitiously crystallising the structure of the intersubjective relationship has been speculative and equally damaging to its antagonist: radical subjectivism. The focus on

intersubjectivity, especially in its Levinasian matrix regarding discourse on the "face of the Other", seems to have led to the hypostatisation of the intersubjective relationship. Not only is the relationship exclusively between subjects, not mediated by anything but instead photographed in a kind of a-space and a-time, but these subjects are themselves emptied of any phenomenological, political, moral, or in short empirical characterisation. The two subjects of the intersubjective relationship have become pure subjects of the relationship of recognition. The ethics of care, in its own forms (perhaps now rhetorical) of "being exposed to the other", "recognising the other", etc., has taken this form of hypostatisation to the extreme. In this sense, the non-human already present in the social world, the concept of sociomorphism, and Simondon's object mediation are all tools for questioning – and overcoming – both subjectivism and the hypostatisation of intersubjectivity.

3 In this regard, we have already mentioned the systemic–relational perspective of Watzlawick et al. (2011) and Palazzoli et al. (1993), wherein basically every interaction is a form of modification of the content conveyed by another.

4 We will not dig into this notion now. Below, we further discuss the concept of the real. What we can say in advance is that "real" for a system is anything the system recognizes as its environment.

5 Among other things, the concept of sociology of the social can be interpreted as a panorama in itself. In this sense, let us recall once again that the act of describing reality is an ordering act and that the very sciences, such as sociology, that set out to describe reality also perform normative acts.

6 From now on, we will mix the terms used by the different authors to show how the concepts expressed refer to very similar reflections.

7 Returning to Latour's language, we would speak of a mediation event that disrupts the chain of intermediations.

8 We have already used this term earlier, but now it is useful to specify its meaning. We use "socio-symbolic structure" to designate that set of socialised narratives and representations that describe events through utterances shared by a certain group of actors. The socio-symbolic construction thus essentially appears as a set of utterances that can be made by a group of actors to signify objects and events in the world. The latter are the "signifiers", i.e. elements of reality that have the potential to be described by a socialised utterance. We will see later that this definition of the socio-symbolic fits well with Maturana's concept of "cognitive domain".

References

Awad, E., Dsouza, S., Kim, R., Schulz, J., Henrich, J., Shariff, A., Bonnefon, J.-F., & Rahwan, I. (2018). The moral machine experiment. *Nature*, *563*(7729), 59–64.

Badiou, A. (2007). *Being and event*. A&C Black.

Bisconti, P., & Carnevale, A. (2022). Alienation and recognition: The Δ phenomenology of the human-social robot interaction. *Techne: Research in Philosophy and Technology*, 26(1), 147–171

Crisi. (1997). *Il Vocabolario Treccani* (Istituto E). Rome

Heerink, M., Kröse, B., Evers, V., & Wielinga, B. (2010). Assessing acceptance of assistive social agent technology by older adults: The Almere model. *International Journal of Social Robotics*, 2(4), 361–375. https://doi.org/10.1007/ s12369-010-0068-5

Henry, B. (2013). Dal Golem ai cyborgs. *Cosmopolis*, 2, 1–10.

Henry, B. (2014). Imaginaries of the global age. "Golem and others" in the post-human condition. *Politica & Societa*, 3(2), 221–246.

Ihde, D. (1990). *Technology and the lifeworld: From garden to earth.* Indiana University Press.

Kachouie, R., Sedighadeli, S., Khosla, R., & Chu, M.-T. (2014). Socially assistive robots in elderly care: A mixed-method systematic literature review. *International Journal of Human–Computer Interaction*, 30(5), 369–393. https://doi.org/ 10.1080/10447318.2013.873278

Latour, B. (2002). Una sociologia senza oggetto? Note sull'interoggettività. *LANDOWSKI E.; MARRONE G. La Società Degli Oggetti. Problemi Di Interoggettività.* Meltemi.

Latour, B. (2005). *Reassembling the social: An introduction to actor-network-theory.* Oxford University Press.

Palazzoli, M. S., Cirillo, S., Selvini, M., & Sorrentino, A. M. (1993). *I giochi psicotici nella famiglia.* Cortina Raffaello.

Peiffer-Smadja, N., & Cohen, L. (2019). The cerebral bases of the bouba-kiki effect. *NeuroImage*, 186, 679–689.

Russo, F. (2022). *Techno-scientific practices: An informational approach.* Rowman & Littlefield.

Seibt, J. (2016). "Integrative social robotics" – A new method paradigm to solve the description problem and the regulation problem? *Frontiers in Artificial Intelligence and Applications*, 290(September), 104–115. https://doi.org/ 10.3233/978-1-61499-708-5-104

Seibt, J. (2017). Towards an ontology of simulated social interaction: Varieties of the "As If" for robots and humans. In *Sociality and normativity for robots* (pp. 11–39). Springer.

Seibt, J., Vestergaard, C., & Damholdt, M. F. (2020). Sociomorphing, not anthropomorphizing: Towards a typology of experienced sociality. In *Frontiers in Artificial Intelligence and Applications* (Vol. 335, pp. 51–67). https://doi.org/ 10.3233/FAIA200900

Sharkey, A., & Sharkey, N. (2012). Granny and the robots: Ethical issues in robot care for the elderly. *Ethics and Information Technology*, 14(1), 27–40. https:// doi.org/10.1007/s10676-010-9234-6

Turkle, S., Taggart, W., Kidd, C. D., & Dasté, O. (2006). Relational artifacts with children and elders: The complexities of cybercompanionship. *Connection Science*, 18(4), 347–361. https://doi.org/10.1080/09540090600868912

Volonté, P. (2017a). Il contributo dell' actor-network theory alla discussione sull' agency degli oggetti. *Politica & Società*, 1, 31–58.

Volonté, P. (2017b). Il contributo dell' actor-network theory alla discussione sull'agency degli oggetti. *Politica & Società*, 6(1), 31–60.

Watzlawick, P., Bavelas, J. B., & Jackson, D. D. (2011). *Pragmatics of human communication: A study of interactional patterns, pathologies and paradoxes.* W.W. Norton.

Žižek, S. (1999). *The ticklish subject: The absent centre of political ontology.* Verso.

4 A systemic theory of socio-symbolic organisations

Actor–network theory (ANT) constructs a system of complex relationships between networks but does not explain why some build and others collapse. The principle of mediation remains obscure on the side of the actor – what does it mean for the actor to be a mediator? – and on the side of the network – what action produces a change in the balance of the networks? From Badiou we received a fundamental suggestion concerning the "Event" that disrupts the state of Being of a certain historical moment: it becomes Truth only when a group of actors recognise it as such, in a retroactive and narrative construction of the action of mediation. The limit of such a constructivist standpoint is that it wraps itself around narratives without linking this process with the real, the material.[1] To argue that a mediating event, which changes the state of a network of associations, occurs only insofar as it is retrospectively re-read by actors as such would lead to a classic conceptual short-circuit of constructivism. If the world is nothing more than a semantic structure arbitrarily produced by human actors and is substantially disconnected from empirical and hyletic data, how did the semantic structure itself come into being, and why is it (mostly) adaptable to the external world? In other words, if reality has no impact whatsoever on symbolic structures, by what forces do these structures organise and modify themselves? A radical constructivist approach always either ends up in an infinite regression of the principle of symbolic modification or ultimately suffers from logical circularity. In short, it attempts to construct a closed, self-justifying system consisting of the semantic production of individuals. In this sense, the most radical drifts of constructivism are, politically, another manifestation of Cartesianism, which elides the "hardness of the real" in favour of human actions. The total obliteration of the world of things, of objects, in the world of language produces, once again, a surreptitious omnipotence of the human with respect to the objectual, by the same mechanism that we highlighted in the introduction to this manuscript. It is therefore necessary, while reintroducing the question

DOI: 10.4324/9781003459798-4

of language, to carefully preserve the material structure of the real, its resistance to symbolisation. While reintroducing the question of meaning means theorising socio-symbolic networks, it is important in doing so to thematise the relationship between these and the pre-symbolic structure, the "not yet symbolised".

The fundamental problem we have identified in ANT is the impossibility of describing why one act mediates the social system while another intermediates it. We aim to answer this question without reintroducing ontological distinctions between human and non-human actors, which implies leaving aside the set of concepts that includes intentionality and consciousness. These concepts, as we explained in previous pages, reinforce the anthropocentric dynamic in the description of the social. Furthermore, it is necessary to keep a sharp eye on not yielding to the lure of an excessively constructivist theory, where the material is definitively elided.

In order to understand the conditions under which sociality is mediated, we should first ask the opposite question: under what conditions is reality intermediated? In fact, the function of the social is primarily that of normalising actors' acts that might otherwise destabilise the system. And this would probably be the role of Latour's "panoramas" – basically the role of ensuring a vision to be shared by the actors so that they remain as faithful as possible to the functioning of the network. We can identify a principle of self-preservation of social networks, which tend to prolong their existence, like a glue that holds a certain group of actors together. If this principle did not hold, we would see complex human social systems endlessly crumbling and rarely settling into a stable configuration. An initial question that moves us forward is: why is it so rare that a mediation capable of undermining a network takes place, be the network small or large, highly or loosely interconnected? Contrary to what Latour (1992) has argued, non-human actors do not seem to exhaust the set of "missing masses" of the social.

Instead, we support the hypothesis that plasma can be described as the set of possible configurations of meaning that a given event can assume within the socio-symbolic organisation of a network of actors. Every action of an actor, every event, every object is in the constant condition of being able to mean something to the other actors in the network. We will therefore describe it with the term *signifier*, meaning a material/experiential entity that can assume a socialised meaning within a network. However, this process of signification is not univocal: the same event, the same experience could be described by multiple statements, each giving it a different signification.[2] What associates networks, puts them together, inclines them to intermediation and therefore stability is the agreement of a number of actors on statements describing classes of phenomena. Within a social system, there are multiple networks – the temporary associations between actors – that compete to establish the signification of a given signifier. This multiplicity represents all the instances competing over the signification of a signifier. This multiplicity also represents all the possible stable configurations of meaning that the signifier may assume for the different networks inside the social system. These stable configurations are different because each association of actors includes that signifier in a different sociosymbolic system, with different shared plug-ins and panoramas. When a signifier shows up within a social network, a multiplicity of competing associations of actors try to enforce their socio-symbolic interpretation of it. This can be described as a competition for socio-symbolic hegemony. Plasma in this sense represents all the socio-symbolic frames that have the potential to become interpretive paradigms (panoramas) for a given signifier within a specific network. This process leads to phenomena of enrolment, and alliance or conflict with other networks of actors carrying out different socio-symbolic paradigms. Most of these paradigms remain latent in the plasma as *enunciative possibilities*. Only when a new network of actors is formed can such an enunciative possibility emerge from the plasma. That is, the new network can then adopt a statement that describes a signifier in a way that is discontinuous with previous sociosymbolic configurations.

The Truth of an Event, its retrospective signification, is thus given by the emergence from the plasma of a statement that disrupts the previous processes of intermediation. The difference between a mediating and an intermediating action therefore essentially consists in the ability of a set of actors (or even one actor, a "spokesperson" in Latour's terms) to query the plasma structure in order to provide a new statement for a signifier that can bring together a new set of actors, more capable than others of accounting for a signifier. Up to this point, however, objects would remain in the background, simply enrolled by human actors in a sort of hegemonic war to impose a certain socio-symbolic structure[3] within the network of networks. Our intention, however, is not to bring ANT back into the fold of an anthropocentric vision, where objects play an exclusively instrumental role. Moreover, it would not be clear why some actors should leave one network to join another, unless we give "intentionality" back to some actors (the human ones) and not to others (the non-human ones). Nor would it be clear on what criteria a signifier can assume one given meaning rather than another for a network. Both these problems can be avoided by describing networks, at this point made up of signifiers-actors,[4] in a systemic way. In this fashion, the relationships between networks do not require "intentionality" of actors: they are instead guided by a principle of internal equilibrium, which takes its cue from Maturana and Varela's theory of autopoiesis and Simondon's principle of metastability.

A systemic theory – Maturana and Varela

In order to construct a systemic theory of socio-symbolic structures, we must refer to a number of authors whose conceptual productions largely inform the theory of socio-technical systems that we are developing. Among these, certainly Maturana is a forerunner, heir to the cybernetics of Norbert Wiener, who with his colleague Varela developed a systemic theory based on the concept of "autopoiesis". In their book *Autopoiesis and Cognition* (Maturana & Varela, 2012), the two Chilean biologists aim to answer the question: "What is common to all living systems, which is why we qualify them as living?". First, they introduce a fundamental distinction between a system's own characteristics and the set of properties attributed to it by an external observer, a novelty within cybernetic and systemic theories at the time of their writing. Any system, understood as a composite unit, produces within itself a series of relations and interactions between its components, the elements of the system. This set of relations and interactions is called its "organisation". The set of "properties" of a system, e.g. a mechanical arm's function of lifting an object, is not part of the domain of the system but pertains to the domain of the observer. What pertains to the domain of the system includes only those internal relations that describe its autopoietic nature, i.e. the self-reflexive organisation of the set of relations and interactions within the system. Therefore, "autopoiesis" refers to a system that reproduces its own balance of relations and interactions between its components. In the authors' words:

> It follows that an autopoietic machine continuously generates and specifies its own organisation through its operation as a system of production of its own components, and does so in an endless turnover of components under conditions of continuous perturbations and compensation for perturbations. Thus, an autopoietic machine is a homeostatic system (or rather a static system of relations) that has its own organisation (defining network of relations) as the fundamental variable that it keeps constant. This must be clearly understood. Each unit has an organisation that can be specified in terms of static or dynamic relationships between elements, processes, or both. Among these possible cases, autopoietic machines are units whose organisation is defined by a particular network of processes (relations) for producing components, the autopoietic network, and not by the components themselves or their static relations.
>
> (Maturana & Varela, 2012, p. 79)

What is essential, and completely new in the framework of reflection current in those years, is the recognition that the living system (namely the autopoietic entity) has as its only essential characteristic the feature of maintaining constant *the relationship between the processes that produce its components* – that is, its organisation. Taken specifically, none of the components is necessary or sufficient to describe the nature of a system, and all the components of a system could change while the system still remained itself by virtue of not changing its autopoietic organisation. In fact, the balance between components could be identical in two systems whose components have not a single trait in common. Therefore, the difference between an animal and a car – that is, between a living being and a non-living one – lies essentially in the fact that the car does not autonomously produce the processes that specify its components, which are instead produced by processes external to it.

The previous specification of the distinction between system and observer becomes even more useful now. In fact, what specifies the organisation of a system is not any of the descriptions made by an observer, since these always remain descriptions produced from the specific "standpoint" of the observer. In these descriptions, it is the observer who attributes properties to a system, including its teleological aspects. The "standpoints" of observers are related to the autopoietic nature of the observers themselves. Their descriptions are to be distinguished from those produced by the system in virtue of its specific organisation. This element explains much of Maturana's gnoseological and epistemological positioning. Particularly representative in this sense is the paper on the relationship, in the frog, between vision and the cognitive apparatus (Lettvin et al., 1959). This paper claims that the frog's perceptual system, even before hyletic signals are processed at the cognitive level, "constructs" the frog's reality in a specific way that is useful for the frog's organisation. In particular, because of the frog's interest in flies, the frog's perceptual system is more "interested" in very fast-moving objects than in slow ones. The key point in this argument is that the frog's *perceptual apparatus itself* already organises the world in a way that is functional for its autopoietic system. The same gnoseological constructivism is also to be found in the very definition of the autopoietic system: objective reality, the real as such, is inaccessible to subjects. In fact, it is always mediated by the need to store data about the world in a way that is functional for the maintenance of the autopoietic organisation. In Maturana and Varela, we find an embodied perspective about the gnoseological limitedness of autopoietic units. In fact, these units experience reality pre-cognitively since it is mediated by their internal system of homeostasis. We should not conclude that, in Maturana and Varela, the real does not have its own ontological nature, that there is no reality "out there". Rather, it is not gnoseologically possible to access the real in a way that has not already been mediated by the homeostatic organisation of the observing[5] system. At the same time, this also means that the real has no direct effect on autopoietic systems; rather, its effects are mediated by the compensation that a system implements for perturbation produced by the real in order to maintain homeostatic equilibrium.[6]

It is perhaps on the basis of this reasoning, although not explicitly so, that Katherine Hayles in her book *How We Became Posthuman* (Hayles, 1999) offers some critical observations on the political implications of Maturana and Varela's theory of autopoiesis. We can summarise them in three points: the monadic solipsism of the autopoietic system, the survival of the liberal subject in the political framework of autopoiesis, and the recursive contradiction inherent in the possibility of describing a system as autopoietic. These three are, in our view, intertwined.

The system conceived by Maturana and Varela has a very strong solipsistic component: the autopoietic units are always busy in the management and production of their own processes, like bachelor machines. Autopoietic units are thought of as a closed, hermetic system, where any "external perturbation" must be compensated for to restore equilibrium between the components. But this clear, very strong distinction between the closed autopoietic system and its environment can itself be considered a description made by an observer, from the partial (and prescriptive) standpoint of their internal system. When Maturana describes the linguistic relationship between monads as a capacity to make language itself the object of coordination between autopoietic units, he supports the possibility of a system's using the linguistic description it has produced of the real as itself a topic for linguistic description. In other words, the autopoietic system can use its very description of the environment as an environment to describe. After all, this is a second-level descriptiveness, where language itself is the object of linguistic description. It is on this basis that Hayles claims that the very closure of the autopoietic system is nothing more than a description by an observer, whose "objective" validity is undermined by Maturana's own gnoseological premises. However, we do not merely want to confront Maturana's theory with a logical short-circuit; rather, we want to test whether this description can account for the social as we observe it. The consequences of this closed system quickly extend to the social and political realm. Maturana writes, in the preface to *Autopoiesis and Cognition*, a small chapter titled "Society and Ethics", in which he describes the implications of his theory for social systems and human behaviour. It is curious that Maturana wrote this part without Varela,[7] who did not agree about these sociological "consequences" and wanted to limit the descriptive scope of autopoietic units strictly to the realm of cognitive biology.

Maturana, in this small chapter on society and language, raises a similar question to the one we are asking in this work, and it is therefore of particular interest for our discussion. We quote Maturana in the introduction of *Autopoiesis and Cognition*:

The characteristic feature of human existence is its occurrence in a linguistic cognitive domain. This domain is constitutively social. However, what is a social system? How is a social system characterised? How do living systems in general, and human beings in particular, participate in the constitution of the social system they integrate? The answers to these questions are central to understanding social dynamics and the process of social change.

(p. xxiv)

However, what is not found in the pages that follow these words is an explanation of the very process of social change. Here is a summary of Maturana's argument. The basic assumption is that social systems are systems in which the components (autopoietic units, in this case implicitly human) realise their autopoiesis, and this realisation is a necessary and sufficient element for constituting a system as social. The social is, for Maturana, an example of *structural coupling*: an action is social as long as it realises the autopoiesis of the participating units.

Consequently, I propose that a collection of autopoietic systems that, through the realisation of their autopoiesis, interact by constituting and integrating a system that operates as the (or a) medium in which they realise their autopoiesis, is indistinguishable from a natural social system.

(p. xxiv)

Given this premise, social consequences emerge, including those captured by Hayles in describing the political implications of Maturana's approach. Social systems become sets of units enclosed within themselves, whose main objective is to preserve their autonomy and "freedom", here understood in the negative sense of being "free from" interference, able to continue to produce their components. Maturana puts great emphasis on the hermetic closure of the autopoietic unit, which has no direct relation to the world because of its gnoseological narrowness at the level of the perceptual apparatus. Here Hayles underlines a return, albeit very advanced, to humanist liberalism. This, in Maturana, is completely self-referential. Nevertheless, Hayles admits, the consequences of autopoiesis for the liberal subject are highly impactful. The structure of self-awareness itself becomes no more than a completely peripheral aspect of the structure of the living being: it in fact represents only a quantitative increase in complexity of the autopoietic system, not a qualitative change. Self-awareness is simply generated by the self-reference of the linguistic system, a sort of screwing of language that begins to describe itself, generating self-awareness with respect to its own linguistic description.

A status report, then: information's body is still contested, the empire of the cyborg is still expanding, and the liberal subject, although more than ever an autonomous individual, is literally losing its mind as the seat of identity.

(Hayles, 1999, p. 149)

In any case, Hayles points out, the structure of the autopoietic unit does not fit with the worldly individual so much as it fits with a radically solipsistic, completely enclosed transliteration of the individual. Language as a means of communication, called *languaging* by Maturana, is nothing more than a *trigger* that activates the other autopoietic units in their self-referral. There is no shared dimension of language but only the position of the observer, who believes that the (teleological) properties of a system belong to the system, while they are a product of its own autopoiesis. Moreover, the main (if not exclusive) use of language on this view is that of coordinating practical activities, a necessity for the structural coupling and autopoiesis of the units (Mingers, 1991). Language, in Maturana, is therefore a false bridge to intersubjectivity from which the autopoietic subject remains structurally excluded.

From the issue of language, and specifically from a complete reconceptualisation of it within Maturana's systemic theory, we believe we can move on in our journey: first to reshape the relationship between humans and the world in a non-anthropocentric form; second, to give an explanation of the processes of modification of social systems, starting from the introduction of relational artefacts. Hayles herself acknowledges an important advance for autopoiesis theory in decentralising the human, while keeping the structures of the liberal autonomous subject firmly in place, indeed reinforced.

Let's move on towards our objective, then. On the one hand, Maturana's description of social systems explains very effectively why social systems are stable. On the other hand, it is very difficult to grasp, within his theory, how they come into crisis. In fact, conservation is given by the fact that all autopoietic units involved in structural coupling need the social system to realise their autopoiesis (for a banal example, hunting in a group brings more food than doing it alone). Maturana explains the homeostasis relationship as follows:

A society, therefore, operates as a homeostatic system that stabilises the relationships that define it as a social system of a particular type.

(p. xxvii)

Each autopoietic unit also self-defines its components and participates in the social system to the extent that this increases the internal stability of the components and the unit's realisation. The role of the social system is therefore to *stabilise*. But how can this monadic process of structural coupling enter into an evolutionary crisis? In the process described by Maturana, the evolution of autopoietic systems is extremely teleological. What happens consists of small processes of continuous settling between the autopoiesis of the units and the social structure as the *medium of* structural coupling. What cannot be explained in this framework are immediate, sudden, and very rapid changes in the whole systemic organisation of societies. The networks of associations, to return to Latour's language, do not follow a tranquil path of uninterrupted modifications and settlements but are rather characterised by an inconstant trend of long periods of immobility interspersed with occasions of sudden modification of most of the structures.[8] Also, even if we could easily explain why systems come into crisis – e.g. structural changes in a system's environment, such as new climatic conditions, famine, or epidemics – we would still fail to understand how autopoietic, completely self-referential units can reconstitute a new form of society so quickly. In fact, when we consider the self-referentiality of autopoietic units, their autonomy and "self-containedness", we cannot understand how they can be re-enrolled so quickly: if the unit was involved in structural coupling within the previous social system, this necessarily means that its internal homeostatic equilibrium was deeply linked to the system's equilibrium. Adherence to another social configuration, that is, to another social medium of structural coupling, requires that the new social configuration be able to involve autopoietic units without demanding an excessive modification of their internal organisation. As Maturana says, autopoietic units participate in a social system *only* to the extent that it realises their autopoiesis. We must therefore necessarily think that no modification of social systems in history has been destabilising enough to prevent a large proportion of the units from adapting their organisation to the new system, obviously an implausible supposition. Or we should think that there is something missing in Maturana's theory as applied to social systems, something that can explain how autopoietic systems are able to reorganise their relationship with a new social configuration so rapidly, breaking to some extent the homeostasis that the theory of autopoiesis describes so well.

The homeostasis of socio-symbolic system

Even if we grant that this move rescues the theory from solipsism, the theory still seriously understates the transformative effects that language has on human subjects.

(Hayles, 1999, p. 147)

Maturana explains why actors in networks are intermediate and why, in most cases, associations of actors are solid and tend not to unravel easily. The swarm of associations, all of which can be regarded as mediums through which actors achieve structural coupling with the environment, remain chains of intermediation for as long as they are able to support the actors' homeostasis. It follows our first conclusion: actor intermediation is thus a consequence of maintaining homeostasis, while the process of mediation is a consequence of losing it.

Previously we described the concept of plasma as the set of possible meanings, of enunciative possibilities, that a given *ens* (object or event) can assume in the socio-symbolic framework of a network of associations. The domain of the socio-symbolic, translated into a systemic dynamic, thus becomes that set of significant paradigms that can generate a homeostatic equilibrium for the autopoiesis' interpretation of the environment. We find this concept in Maturana, although expressed differently:

> The domain of all interactions into which an autopoietic system can enter without losing its identity is its cognitive domain; or, in other words, the cognitive domain of an autopoietic system is the domain of all the descriptions it can make.
>
> (Maturana & Varela, 2012, p. 119)

The cognitive domain of an autopoietic unit is thus constructed from the set of possible, and therefore meaningful, descriptions that it can generate about the environment. Each particular network of association, which enrols objects and events to constitute a meaningful landscape, is a specific instance of the possible relations between signifiers and meaning that the cognitive domain can produce to signify events. To break down this definition, the "specific instance of possible relations" is the particular equilibrium that constitutes the cognitive domain of that given autopoietic system, be it a social system or a single actor. It is that given configuration of the socio-symbolic that has emerged from the relationship between all the possible statements that that social system might produce (the plasma) and the external environment. The set of signifiers comprises the elements of the environment that are in the condition of being signified by the autopoietic system, meaning that they can at least be perceived.[9] This means that the possibilities for homeostatic equilibria of socio-symbolic systems are always more than one. Given any number of signifiers – actors, objects, and events that can in principle be significant in the cognitive domain of an autopoietic unit – there are an almost infinite number of possible homeostatic equilibria that might serve as a stable socio-symbolic frame. Let us differentiate between the concepts of plasma and socio-symbolic framework. The first constitutes all the possible statements that a social system as a whole can enrol to give meaning to the environment. Most of these statements cannot be enrolled simultaneously by the system to signify the environment, because to enrol all of them at once would produce contradictions and violate the internal homeostatic equilibrium. For example, a social system cannot allow the statements "a free market improves economies" and "a free market worsens economies" to signify events at the same time, because this would entail a contradiction. Therefore, the socio-symbolic organisation enrols only a partial selection of all the possible statements in the plasma, because only some of them can be included in a symbolic system at any given time without self-contradiction. Moreover, since every socio-symbolic system must enrol actors to form a panorama, it must signify the environment in such a way that actors will enrol in that association. Therefore, the socio-symbolic can also be described as the lowest common denominator of all the cognitive domains of the autopoietic units participating in an association. From this point of view, the plug-ins and panoramas described by Latour no longer simply become "packages" to be stochastically downloaded by an actor in order to enrol in a specific network. Now an actor joins a network because the relationship between signifiers that it expresses is positive for the actor's homeostatic equilibrium – that is, it allows structural coupling.

In this process, objects have been neither surreptitiously excluded from the social nor given a mimetic anthropomorphic agency of intentional processes. Actors are still, as in Latour, enrolled by autopoietic networks of associations, and their mediating principle is to be signifiers that, sometimes unexpectedly, upset the balance of cognitive domains (at the unit level or the social level), bringing from the background to the proscenium relations of signification that had previously remained in the plasma as pure possibilities of signification.[10]

The socio-symbolic system can be in turn understood as an autopoietic system consisting of the set of cognitive domains in which the units/actors participate. The associations of actors thus become instantiations of the socio-symbolic, which includes in itself all the existing systems of relations between signifiers and all the panoramas expressed by the associations in the network.

Let us now summarise what has been said so far. Socio-symbolic systems are made up of relations between meanings and signifiers and correspond to Maturana's cognitive domains, constituting the panoramas of associations. Included in the plasma are all the possible statements that might signify the environment but are in the background. A statement will either remain silent in the plasma or actively signify the environment according to two rules of homeostasis: (i) the socio-symbolic system itself can enrol new statements to signify reality only under certain systemic conditions (Bisconti & Cesaroni, Forthcoming) – for example, the statements must jointly meet a condition of non-contradiction; (ii) actors subscribe to one system of signification rather than another on the basis of the greatest possible internal balance. This explains why, among other things, an actor can enrol (even if only partially) in several networks, since each actor builds a personal set of statements interpreting the environment.[11] Therefore, we are not abstracting the actor, as if the actor were a disembodied hypostasis. Networks of associations are simultaneously made up of human and non-human actors, with no phenomenological difference in their participation in the network: both may either mediate or intermediate the socio-symbolic system on the basis of being signifiers meaningful for other actors. Indeed, each actor, whether silicon-based or carbon-based, contributes to shaping the socio-symbolic organisation and its autopoiesis, including by means of new signifying statements that were previously in the plasma. The actors all contribute to the systemic equilibrium of the socio-symbolic system with which they are coupled. Obviously, a single actor is rarely able to modify the socio-symbolic organisation alone by enrolling paradigms from the plasma. However, an association of actors might have the "critical mass" to modify the socio-symbolic organisation. This systemic vision allows us to entertain the possibility that even an immobile actor, who does not act at all, can mediate the system: the continuous modification of the relations of meaning in the socio-symbolic system can place a completely inert actor in the position of producing a systemic disequilibrium. In fact, if the paradigm changes, what was previously not producing a crisis might no longer fit with the new paradigm, ultimately unbalancing the system. In the never-ending reshaping of the socio-symbolic structure, it does not matter whether the cause of the disequilibrium is a human actor or a non-human one.

The crisis of socio-symbolic systems

> At length I remembered the last resort of a great princess who, when told that the peasants had no bread, replied: "Then let them eat brioches".
> (Jean-Jacques Rousseau, Confessions)

Nevertheless, something still seems to be missing within the systemic equilibrium of the socio-symbolic that we discussed in the previous section. The system is still completely self-referential by way of the autopoietic units, rendering it completely detached from the outer world. Even though these processes are now included in a laborious production of signifiers shared between actors, autopoietic units are producing equilibria in relation only to themselves. Moreover, it is not clear how symbolic production can generate equilibrium in systems. As Hayles' critique indicates, it seems that Maturana radically underestimates the importance of language, reducing it to a mere instrument of cooperation between individuals. Why does language seem to be so essentially at the root of actors' social constructions and the socio-symbolic crises they trigger? To address this question, it is now necessary to investigate the relationship between the real and the symbol. In the introduction to this chapter, we commented on the secondary role to which idealistic philosophy has relegated the real, the material outside the cognitive domain of the human being. The autopoietic unit is completely enclosed within itself, its perceptual apparatuses only capable of picking up signals that are useful for the realisation of autopoiesis; yet this seems to explain only a small part of worldly experience. Living beings are constantly in a position to experience events that are traumatic for the organisational structure of the actor. Thus, social systems also come into crisis in an unexpected way, suddenly exploding and recomposing themselves into new equilibria that previously seemed impossible. What makes this process so sudden? If reality were faithful to the theories of Maturana and Varela, the possibility of sudden changes would not arise. Great historical events of social restructuring would be impossible because the autopoietic units would migrate into a new structural coupling as similar as possible to the previous one. This would certainly be less expensive than any crisis in terms of homeostatic readjustment. In the previous chapter, we gave an account of the autopoietic organisation of socio-symbolic structures, which organise social systems – but only in the case of intermediation. What is at stake, then, is to understand what process sometimes drives networks of actors to blatantly violate homeostatic and autopoietic principles, enrolling from the plasma a Truth/Event that completely disrupts the previous socio-symbolic equilibrium and suddenly creates a completely new one.[12]

To explain this phenomenon, we will use Simondon's (2020) concept of metastability. As opposed to Maturana's teleological, tranquil, and linear procedures of systems' rebalancing, metastability allows us to think of a system that possesses a certain potential energy that allows for sudden modification, which remains covert until the modification occurs. Every socio-symbolic system, described as a cognitive domain, can produce a finite set of descriptions that number as many as the possible relations between signifiers that can be expressed by that system. This means that the socio-symbolic system bases all its expressive capacity on the number of signifiers actually experienced, not on all those that can be experienced, and on all the relations actually expressed between the signifiers, not on all those that can be expressed. The occurrence of an event that is not yet described in the cognitive system necessarily brings about a certain imbalance within that system as it (or the socio-symbolic system, if we consider social systems) must include that new event among the meaningful experiences it recognises. To include a new signifier implies a temporary imbalance since every addition to the cognitive system requires a systemic change. A basic rule of systems theory is that every event modifying a part of the system has implications for the entire system (Bateson, 2000). Accordingly, the disequilibrium introduced by the new event increases the metastable energy of the system. In fact, since homeostasis is essentially a conservative principle, the system will not modify itself by enrolling new signifiers or modifying the relations between signifiers every time an event occurs that cannot be included in the socio-symbolic structure. For every time a system must include a new enunciative structure from the plasma in order to signify the real, it must bear a certain "cognitive cost" in carrying out this operation. The cognitive cost is the effort of producing a new enunciative-systemic balance of the cognitive system, or the socio-symbolic structure if we focus on social systems. As post-rationalist psychology reports (Guidano et al., 2019), a whole series of strategies can be implemented that allow a system to limit the impact of an event within the individual's structure (exemplified in the concept of *coalitional control*, for example). One of these strategies, as we have seen above, is to minimise the disequilibrium that an event can bring to the socio-symbolic structure of a system. The plasticity of the enunciative structures allows the socio-symbolic system not to be modified every time an unforeseen event happens. Each of these strategies preserves the system from modification and maintains its homeostasis. In fact, these strategies allow the system to tolerate new events without producing actual changes in its organisation.

Let's further explain this crucial point, particularly since our claim here is different from Maturana's. For Maturana, the autopoietic unit seeks homeostatic equilibrium, changing its organisation in relation to the environment. This entails a certain effort and therefore a cognitive cost for restructuring the equilibrium. In contrast, adopting the concept of metastability allows for the possibility that a system can avoid changing the socio-symbolic system it uses to describe reality. In fact, the system can employ different strategies, such as plasticity and partialisation, to include a new event inside the current socio-symbolic organisation. At the same time, this process increases the entropy/energy in the system, making it metastable. This increase in energy, and therefore in metastability, produces a progressive detachment between the real/the environment on the one hand and the socio-symbolic structures supposed to explain it on the other. The socio-symbolic system becomes progressively less adapted to the events, and ultimately the environment, that it is supposed to explain. This separation eventually leads to a rupture, which may be caused by an insignificant event or one that is not in itself particularly more destabilising to the system than all the previous ones. The accumulation of energy brings the metastable system to a breaking point.

How does metastability work? What allows a system *not* to change, notwithstanding the fact that the external world is actually changing, and therefore the environmental fit is worsening? Here, the strategies that the socio-symbolic system (or the cognitive domain, on the individual scale) can employ come into play. When employing strategies to avoid changing in relation to a changing environment, the system is making a "choice": to increase the metastability, a prospective imbalance in the system that can lead to crises, instead of paying a cognitive cost for the effort of reshaping the socio-symbolic system (namely the relation between statements and paradigms) and ensuring environmental fitness. Why should a system make this choice? Because in the relationship with the environment there is a degree of uncertainty: an association of actors producing socio-symbolic systems cannot predict with accuracy whether the change in the environment is actually a modification capable of launching the current socio-symbolic system into a crisis. Therefore, not every change in the environment is worth the cognitive effort of reshaping the socio-symbolic organisation. The choice between paying a cognitive cost and increasing metastability is a bet facing every socio-symbolic system. The strategies that enable a system to increase its metastability and avoid systemic change are well explained by Guidano (1992) and Guidano et al. (2019) from the perspective of post-rationalist cognitive psychology, which owes much to Maturana and Varela. We take inspiration from this field to summarise the strategies implicit in the concept of the *plasticity* of socio-symbolic systems.

Plasticity is a property of socio-symbolic systems implying that statements enrolled from the plasma and constituting the socio-symbolic must be adaptable and "malleable" enough to explain an ever-changing environment. In fact, any interpretive system, in order to be effective, must be able to articulate itself around contradictory manifestations of the environment. The plasticity of a socio-symbolic paradigm (a set of statements, namely a panorama) allows a system not to enrol new statements from the plasma every time there is a modification of the environment. Instead, the system can change the relationships between existing statements or otherwise adapt them to the new configuration of the real. In order for a statement to be plastic, it must be able to describe a certain number of events in the real. The more events it can describe, the more plastic it is. For example, the classic Marxist statement "History advances because of the struggle between classes" can interpret a very large variety of events. So can the classic capitalist statement "The market will grow endlessly; when it falls, that is only a parenthesis", which has the same level of abstractness, allowing it to give meaning to nearly any event related to the economy. Notice that these two statements can enrol highly contradicting event/signifiers. These are examples of very plastic statements that have shown the ability to enrol vast numbers of actors inside their respective panoramas, being suitable for most autopoietic organisations because they can bear contradiction (in this case, we could call these panoramas "ideologies"). A paradigmatic system, namely a set of statements interpreting a class of events, must be capable of being reused to interpret different environmental configurations pertaining to the same interpretive domain (e.g. economics or international politics). Thus, the more plastic a paradigmatic system is, the more adaptable it is without modification. This characteristic of socio-symbolic systems usually implies the production of paradigmatic systems composed of statements at a rather high level of abstraction. The risk of excessive system plasticity is, of course, that paradigms become overly abstracted and empty interpretive structures, to the point that they are no longer effective for interpreting the environment.

Another fundamental principle of socio-symbolic systems is the principle of economy with respect to paradigmatic structures. The proliferation of paradigmatic structures requires a twofold effort in order for the subject to maintain a stable cognitive structure for interpreting the real. First, the proliferation of statements requires the autopoietic unit to exert cognitive effort in discerning which paradigm is adaptable to which event. Second, the existence of numerous paradigms creates faults, constant ruptures in the system of interpretation, since every event is usually simultaneously interpretable by several paradigms, often mutually contradictory, requiring a cognitive effort to order this multiplicity. Following the principle of economy as applied to socio-symbolic systems, then, the environment should be explained by means of as few statements as possible so as to create a maximally stable, reliable representation for interpreting the world.

The principle of economy goes hand in hand with that of plasticity. The more plastic a statement is, the more economical it is.

The more plastic a system's statements are, the more metastable the system becomes, since the fit with the environment (namely the fit between interpretive paradigms and signifiers) correspondingly decreases. At some point, the socio-symbolic system could fail to adapt to a novel configuration of the environment. Given this high level of ineffectiveness due to excessive metastability (due in turn to excessive paradigm plasticity), a new association of actors might then be able to mould itself around a new socio-symbolic configuration, enrolling statements previously contained in the Latourian plasma. This new association, promoting a new socio-symbolic organisation, could then suddenly overpower the old one and become hegemonic in the social system.

If accepted, this reasoning finally allows us to account for the problem that Latour identified:

Why do fierce armies disappear in a week? Why do whole empires like the Soviet one vanish in a few months? Why do companies who cover the world go bankrupt after their next quarterly report? Why do the same companies, in less than two semesters, jump from being deep in the red to showing a massive profit? Why is it that quiet citizens turn into revolutionary crowds or that grim mass rallies break down into a joyous crowd of free citizens? Why is it that some dull individual is suddenly moved into action by an obscure piece of news? Why is it that such a stale academic musician is suddenly seized by the most daring rhythms? Generals, editorialists, managers, observers, moralists often say that those sudden changes have a soft impalpable liquid quality about them. That's exactly the etymology of plasma.

(Latour, 2005, p. 245)

The reason for the abruptness of historical occurrences with regard to social systems is the accumulation of energy that makes social systems structurally metastable. Plasma, namely the set of relations between unexpressed and latent signifiers in a system, can suddenly destroy the previous networks of associations, disrupting in a few moments the previous calm flow of social intermediations. Metastability can be therefore described as a state of system stress that presents no visible effects until the system suddenly collapses, reaching a breaking point at which another socio-symbolic organisation is preferable. This means that at this point, the actors prefer to pay the cognitive cost of a systemic change because the fit with the environment has become too low and the socio-symbolic system is no longer able to reliably describe the environment. We have therefore finally come to define Latour's principle of mediation: a mediator is an actor who, in relationship with the other signifiers of the system, produces an excess of energy in the existing socio-symbolic organisation and suddenly breaks it up. We have also come to define the principle that governs the enrolment of a series of actors in a Truth of the Event, which upsets the plane of Being. The process of a system's growing metastability leads some of its actors to be progressively less and less in equilibrium within the socio-symbolic organisation (the Being, the panorama), namely less and less in structural coupling as the social system achieves diminishing levels of autopoiesis. The inexpressibility of this disequilibrium[13] progressively increases metastability to a point of total rupture with the previous system. Metastability also explains why Latour's goal of observing the continuous mediation events that actors perform is simple at the micro level, while those events yet remain invisible at the macro level. If the mediating event represents a certain crisis for the system, perhaps a microscopic one, this does not necessarily change the socio-symbolic equilibrium. In most cases, it serves only to increase its metastability. We should understand metastability as the degree of elastic tension that exists between the configuration of the actual relations between actors in the world and the symbolic structures (panoramas) supposed to explain these configurations. If the elastic tension is very high, i.e. the system is highly metastable, then the symbolic structure supposed to signify that set of relations is less effective than it used to be. Moreover, as long as the excessive metastability does not break up the current socio-symbolic configuration, the fact that a crisis is coming may be invisible. The crisis will happen when a network, or a spokesperson for it, is able to enrol a meaningful paradigm (a panorama) from the plasma that unravels the previous socio-symbolic configuration. The new socio-symbolic organisation will be better adapted to explaining the environment: objects, experiences, and relations among actors. This does not necessarily happen easily.

So far, we have defined the process that governs the sudden breakdown of social systems, outlining the differences between mediation and intermediation in socio-symbolic systems. The principle that allows a set of actors who have broken up a previous network to reform into a new one just as quickly remains obscure. If we were to follow Latour here, we should expect the principle of creation of the new network to follow the enrolment, more or less stochastic, of new plug-ins and panoramas. This would be incomprehensible, however, because the reorganisation in fact occurs in a synergic and concerted manner. If, instead, we credit Maturana's view, we should expect the autopoietic units to look rationally for a system as close as possible to the previous one, with which to engage in structural coupling in order to maintain homeostatic equilibrium as much as possible. In fact, the autopoietic units should always choose not to waste energy and prefer to maintain the current autopoietic organisation. None of these solutions seems to account for the processes that regulate the incredibly rapid formation of new, solid networks, sometimes strongly different from their predecessors. Therefore, we have not understood why autopoietic units accept the waste of the enormous amount of cognitive energy required to completely reshape the socio-symbolic system, instead of adopting a system that allows better environmental fit with the least possible cognitive effort.

What follows is a proposal for an explanation, which will certainly not close the discussion. Nevertheless, having come this far, it seems necessary to venture into this last theoretical challenge.

The problem of the formation of new networks remains obscure until the moment when a fictitious functional distinction is made between the universe of the real and the universe of the symbolic. Following a rigid distinction, the real – the ὕλη – would not have any effect on a subject if not in the form of something that has become significant for the subject. At the same time, the symbolic could not exist if not in constant reference to a real, to which it represents the reference. In the creation of a new network of actors, the presence in the real of an element that can stand as an organisational signifier for the new network would therefore entail a challenge. The element should actually exist out there and should be signified in the same way by a large number of autopoietic units. In this way, it could become a medium strong enough to activate the structural coupling of other units. As already discussed, it is the telluric movement of the real that triggers the constant need for the socio-symbolic system's redefinition. The real has a strong capacity for autonomy with respect to the symbolic and manifests the "void at the centre" of the symbolic, to echo Badiou's discourse on Truth/Event. As an internal gap in the socio-symbolic equilibrium, the real does not manifest itself in the form of a meaningful presence. Instead, it underlines the gaps in the system that is supposed to signify it. We claim that the manifestation of the real is first and foremost as a negative referent since its emergence points out the insufficiency of the symbolic. It refers to the absence of something: a signifier that, if it were present, would bring balance to the socio-symbolic organisation. However, this absence is itself a signifier. This is what Hayles would define as the dialectic of presence/absence, which characterises the Lacanian approach to the relationship between the real and the symbolic, namely the procedure underlying the process of *meaning-making*.

> Thus, for Lacan, a doubly reinforced absence is at the core of signification – the absence of signifieds as things-in-themselves as well as the absence of stable correspondences between signifiers. The catastrophe in psycholinguistic development corresponding to this absence in signification is castration, the moment when the (male) subject symbolically confronts the realisation that subjectivity, like language, is founded on absence.
>
> (Hayles, 1999, p. 31)

We do not fully agree with Hayles. Actually, we claim that in Lacan linguistic production is found not in the absence of a stable correspondence between signifiers,[14] but in the relationship between the universe of desire (at the root of which *Das Ding* exists) and the flickering of signifiers. In any case, the dialectic of presence/absence is certainly fundamental in the constitution of language. In order to solve the problem of the sudden association of new networks, we will therefore use this dialectic, which is typical of the posthuman way of considering information.

More than anywhere else, the presence/absence dialectic can be found in the concept of the signifier as "covering a trace" (Lacan, 2014). Covering a trace means doubling the absence of the symbol, since the very fact that something is no longer – or not yet – traceable in the real can become significant. In Lacan, the presence/absence dialectic accounts for the fact that the signifier can also constitute itself in relation to an absence in the real, not only to a presence. Hayles is right when she says that for Lacan, "Language is not a code, because he wanted to deny one-to-one correspondence between the Signifier and the signified" (p. 30). But Lacan's position is actually even more radical than this, since the correspondence is not only biunivocal – it can even be reversed in polarity. Not only can the process of signification leverage all the possible configurations of the real and of the relations between its elements, but it can also begin from the absence of a reference to the real, in the form of covering a trace. The metaphor underlying this concept of the covered trace is as follows: someone who leaves traces while walking posits elements that are significant; likewise, someone who erases traces also posits significant elements, since the fact that the trace *is not found there* is a significant element. The signifier that acquires its meaning from the very absence of a reference in the real corresponds to Badiou's idea of Truth, a void at the centre of the Being. In this case, the absence is that exemplified by a socio-symbolic equilibrium with which those who are enrolled by a certain Truth do not find themselves in structural coupling. We believe that the meaningfulness of absence is one of the most radical of Lacan's conclusions with regard to the problem of language. The human being is in fact that very particular animal that can point out that something is not there, that it is not present, and make this a meaningful element.

Hayles also proposes another dialectic: that between *pattern* and *randomness*. The latter manifests itself above all in the process of mutation:

> We can now understand mutation in more fundamental terms. Mutation is crucial because it names the bifurcation point at which the interplay between pattern and randomness causes the system to evolve in a new direction. Mutation implies both the replication of pattern – the morphological standard against which it can be measured and understood as a mutation – and the interjection of randomness – the variations that mark it as a deviation so decisive it can no longer be assimilated into the same.
>
> (Hayles, 1999, p. 33)

The process of mutation, which follows the principles of *information technology*, sheds light on the circulation of signifiers. This circulation can be understood as a replication of patterns of signifiers within the meaning-making process. The meaning-making is due to the possible slippages in this circulation, which in turn are due to the randomness of the process of acquiring the signifier as the referent of an absence. The principle of randomness, allowing mutation in the process of replication (Hayles, 1993), is possible only because there is no necessary connection between symbol and real referent.

Foregrounding pattern and randomness, information technologies operate within a realm in which the signifier is open to a rich internal play of difference. In informatics, the signifier can no longer be understood as a single marker, for example an ink mark on a page. Rather it exists as a flexible chain of markers bound together by the arbitrary relations specified by the relevant codes.

(Hayles, 1999, p. 31)

In our opinion, the concept of randomness also stems from Jacques Lacan. In Seminar VI (Lacan et al., 2019), Lacan delves into the linguistic play of a child, which, for him, epitomises the intricate relationship between the subject, signifier, and meaning (pp. 198–202). The French psychoanalyst examines a child's linguistic behaviour, where the child substitutes the word "dog" with the expression "bowwow". For Lacan, this act of replacing one signifier with another stands as a foundational linguistic operation. The principle in Lacan is: to generate meaning, one signifier must be substituted for another.

Following this initial act, which Lacan terms "metaphorical substitution", the signifier "bowwow" extends to encompass a range of objects, some of which might bear no relation to a dog. This evolution underscores the transition of the signifier "bowwow" from its substitutive role to its emergence as an independent signifier. The statement "the dog goes bowwow" mirrors a straightforward imitative link to reality. However, the moment the game of exchanging "bowwow" with other signifiers begins, it transforms into a truly meaningful utterance.

Lacan identifies three distinct phases in the child's relationship with the signifier "bowwow":

- "The dog goes bowwow": This represents the predicative phase, mirroring reality.
- "The Bowwow": Here, one signifier is replaced by another, serving a referential purpose.
- "Object X = bowwow": The child shifts the signifier "bowwow" from its attributive or substitutive role, bestowing upon it inherent significance.

For Lacan, the act of substituting "bowwow" for "dog" is metaphorical in nature. Yet, at some juncture, the metaphor detaches from its original referent, acquiring significance in its own right. Meaning, in Lacan's perspective, emerges when metaphorical content is generated and subsequently detaches from preceding signifiers.

Consequently, "bowwow" evolves, transcending its initial role as a mere attribute. Meaning, according to Lacan, springs from the act of replacing one signifier with another. This substitution entails a loss – the predicative essence of "bowwow". Nevertheless, this process elevates the signifier "bowwow" to a position of complete autonomy, severing its ties to its original role as a dog's sound. Lacan describes this transformative journey as the mechanism by which the subject navigates the intricate web of signifiers.

In simpler words, these concepts brought together by Lacan and Hayles can help us in explaining why a network can quickly aggregate around a new signifier that is sufficiently shared to create a new, stable association. If such a signifier can be created from an absence in the real, this opens up the arbitrariness of the relationship between sign and reference. This arbitrariness works as in information technology: by way of a relationship between *pattern* – the replication of the signifier in a network of actors – and *randomness* – the continuous *shift* in meaning that can occur in a sign that refers to an absence instead of a presence. In this way, the same signifier without a reference in the real can be enrolled by different autopoieses for different reasons. In fact, the signifier is not pointing towards the environment but only towards a lack in the autopoietic organisation. Even if each autopoietic unit has a different locus of lack, something we might call the homeostatic breaking point, the same signifier can enrol the different units. In fact, this signifier has no positive content (reference to the environment) but only a negative one: it points out that the current autopoietic organisation is not properly signifying the environment. The signifier that refers to an absence and circulates through mutation becomes an element able to balance an extremely large number of autopoietic units.

This, in our opinion, explains the ability of eschatological paradigms to bring together a considerable number of actors in a network. And this is why eschatological contents are almost always to be found in associations of actors that emerge after a destructive socio-symbolic crisis, such as a revolution. In the wake of a socio-symbolic crisis, all the various autopoietic units are in search of a new homeostatic equilibrium and a new structural coupling with a socio-symbolic organisation. While it is very hard to organise a wide, stable network of actors on the basis of a symbolic paradigm with a reference in the real, actors can easily be enrolled by eschatological contents since these can be utilised by each autopoietic unit for its own scopes while sharing in the same socio-symbolic equilibrium. At the core of eschatology, as far as linguistic structure is concerned, is the fact that the eschatological symbol is not related to a real referent but to its absence. The process of mutation, given by the *pattern/randomness* dialectic, explains how this reference to absence can serve as a homeostatic element for a very large number of autopoietic units.

On Haraway's compostist theory

Our autopoietic units now seem much less "autonomous" and closed in themselves than they were within Maturana's theory. The development of their autopoiesis is now understood in structural relation to the real, the outer environment, since elements of the units' homeostasis are now serving as signifiers in a union of enunciative paradigms and materiality. These elements are structurally socialised, whether created in the absence or the presence of the real. Autopoietic units associate in networks because the signifiers, which circulate according to a principle of mutation, aggregate the units. The interweaving of the socio-symbolic, itself a homeostatic system, is the true plane of existence of social actors. On this plane, the events that occur in reality – or even their non-occurrence – become signifiers capable of modifying the equilibrium of the network. Both humans and non-humans participate in these networks, which include not only objects but also mythologies and in general the elements that populate the social imaginary. In this way, we resolve the problem (Volonté, 2017) posed regarding the agency of imaginary actors. Each of these is an "agent" in the networks of associations insofar as it is an element that brings about a modification of the socio-symbolic structure and consequently of the actors' associations. Within Maturana's observer theory, we can say that an autopoietic system is always structurally an observer of itself, as it is only a co-constructor of language. Therefore, the actor has always been split up, has always to some extent looked at itself from outside. It has always "spoken to itself from outside". We believe that Mikhail Bakhtin's concept of dialogism can describe this new unit: each speaker deals with words that are already inhabited by the voices of others, penetrated by the intentions of others. The speakers will always speak with others' words even if, by placing them in their own context, they imbue these words with their own intentions (Holquist, 2003). The speakers will then enter into a dialectical relationship with words that are not neutral objects but have always been the others' words, filled with their meanings. Every speaker's discourse is therefore a movement of approaching and a distancing from the words of others. The linguistic signifier is already loaded with socio-symbolic contents; the autopoietic unity is therefore already structurally contaminated with the outer. And it is not exclusively contaminated by the human other, since humans are no longer the only speakers: every object is a speaker-signifier insofar as, by entering into a relationship with the socio-symbolic homeostatic system, it modifies the system's equilibrium to some extent. Thus, the mere existence of the object, or even the possibility of imagining a certain object, is already a form of hybridisation. The autopoietic unit is now much less impermeable and self-centred than expected, but at this point it is no longer solely autopoietic either: it can no longer constitute itself, nor are its actions in themselves worth anything. The production and circulation of a signifier becomes an element of a system not only in which the unit participates but also by which the unit is structurally constituted. At this point we might accept Donna Haraway's proposal in *Staying with the Trouble* (Haraway, 2016) that we should speak not of autopoietic units, but of *sympoietic* ones. These are shared configurations that, going beyond the principle of the supposed self-sufficiency of living systems, consider transversal processes to be at the basis of evolution, structurally open to otherness. The unity is no longer autopoietic because that αὐτός has lost all reference, both with respect to the self-reference of its own processes and with respect to its self-production.

Haraway's concept of sympoiesis allows us to devote a little space to discussing the extent to which this manuscript is distant from the latest posthumanist theories, which we will represent here with Haraway's (2016) essay.

> Critters are at stake in each other in every mixing and turning of the terran compost pile. We are compost, not posthuman; we inhabit the humusities, not the humanities. Philosophically and materially, I am a compostist, not a posthumanist. Critters – human and not – become-with each other, compose and decompose each other, in every scale and register of time and stuff in sympoietic tangling, in ecological evolutionary developmental earthly worlding and unworlding.
>
> (Haraway, 2016, p. 97)

In this essay, Haraway even goes beyond the notion of the posthuman, embracing that of "compost" as a replacement for it. This concept takes to the extreme the consequences of the deconstruction of the alleged human uniqueness, the alleged human separation from the animal/vegetal/mineral world, in order to mount a definitive attack on anthropocentrism. The Chthulucene, far from being a reference to Lovecraft's monster, refers to the concatenations between human, non-human, and humus and to the generativity of sympoietic processes. In comparison with such "compostist" thought, there is no doubt that this manuscript still gives the human great importance. Dissolving the relevance of humanity within the complexity of the crystal, in the midst of the sprawling and heterogeneous configuration of bodies and world, is a theoretical operation that does not belong to this manuscript and could not in our opinion belong to it. Here we explain two points that illustrate why.

First, this book focuses on understanding how social systems can be modified by the introduction of interacting objects, ultimately producing a normative reflection on the design strategies for social robots. This objective is closely related to that structure called "human society", which according to Haraway is surreptitiously separated from others. Placing objects in their rightful place in the processes of the production, mediation, and destruction of human societies does not mean thinking of them as completely exploded within much larger configurations such as Haraway's *holoent*. We believe that Haraway's approach has two limitations if applied to the objective of this manuscript: surely raising the focus as much as possible from the human being to the sympoietic, "compost" nature of the real has the merit of deconstructing, perhaps definitively, the anthropocentric standpoint. By the same token, however, this approach drastically sacrifices the ability to be theoretically effective in analysing questions that, whether we like it or not, concern human societies. The notions of holoent and compostile remain extremely distant from the implications that a social robot might bring to the interaction between two or more individuals or the normative implications of robot design. The deconstruction of anthropocentrism and the shift from the notion of anthropomorphic interaction to that of hybrid systems have immediate theoretical repercussions for the social system hybridised by talking machines. In contrast, with the concept of *humusity* social relation/interaction takes on a completely different meaning as even the relations/interactions occurring between crystals then count as *social*. In a joke, this is the night where all the cows are social. The social actor him/her/them/itself, whether human or non-human, can no longer be identified and circumscribed within the compost. Beyond a systemic theory, where co-implication between entities is interactional, in the compost approach co-implication seems to be ontological. Haraway's operation is then no longer a questioning of the autonomy of the autopoietic unit, placing it already in a network of interactions, but an attack on the very principle of individuation of the actor. We want to stress that this is not a criticism of Haraway's approach *in se*. We are concerned with the possible field of application of a thought that has moved the theoretical focus so far away from the world of humans that it might be no longer able to deal with that world. It is difficult to imagine the political contribution of a thought whose relationship with the world of humans is beginning to fray. The question we are asking is whether a vision that is so holistic, that reconfigures the way of understanding the world so beyond the humane, is not actually defusing its own political instances. In this theoretical horizon, reconfiguring the design principles that regulate the aesthetics of the social robot, for example in order to avoid sexist dynamics in gendering machines, is an irrelevant position. Indeed, it would instead be necessary to first reinterpret the entire ontological status of the subjects, to emphasise the co-existence and inseparability of carbon- and silicon-based beings. Our concern is that this approach is politically ineffective. The transition from the Anthropocene, as a critical posture of understanding human centrality in the world, to the Chthulucene, a compostist holism of the real, can certainly – on the one hand – be faithful to the most extreme consequences of posthuman thought. On the other hand, it risks losing any political grip on human societies. As in the case of object-oriented ontology (OOO), posthumanism brings the consequences of an ontological position to the extreme, namely the definitive deconstruction of the anthropocentric posture. In doing so, we believe it self-defuses the critical potential that posthumanism had in the political philosophy. Haraway's position is partly descriptive and partly prescriptive, but it is part of a philosophical direction that is used to programmatically take into account its political implications, namely posthumanism. Sadly, it seems to us that the "compostist" approach will be far more effective in analysing the power struggle and the political inequalities between carbon and silicon, than the ones existing in human societies.

We opened this parenthesis on Haraway to make clear why, in speaking of anti-anthropocentrism, we have not taken on board the more radical theoretical consequences of this approach. We believe that the full participation of objects, both material and immaterial, within the human social world is a goal achieved by the first part of our work. Now that we have faced the *macro* level of human–machine interactions, namely how a socio-technical system can be changed by non-human actors, we will outline some political consequences of our reflection.

Notes

1 We are aware that it would not be at all correct to accuse Badiou of radical constructivism; the argument applies only to the consequences of what has been said about the concept of interpretive intervention with respect to a theory of mediation in social systems.

2 What we claim here, in Latour's terms, is that a given plug-in is "downloaded" by an actor because that precise signifier can assume a meaning for the actor that maximises his/her fitness within the social system. This concept of fitness will be explored in the following pages using the concept of "structural coupling", borrowed from Maturana and Varela.

3 The socio-symbolic structure, as explained above, is a system of statements able to give an account of the set of socially signifiable (that can be signified) experiences.

4 In fact, actors themselves are now object of experience that can take on meaning for other actors and therefore are signifiers.

5 It should be noted that this position is already expressed by Lacan (2013) in his commentary on Freud's φ and ψ system. Indeed, the similarities between the gnoseological position of Maturana and Varela and that of the French

psychoanalyst are very strong. We will discuss this a few pages below, when we turn to the problem of the Lacanian Real.

6　A living system is autopoietic in that it is self-producing: it cannot be characterised in terms of inputs and outputs, none of its transformations can be explained as a function of stimuli from its environment; it modifies itself according to its organisation, in order to keep its organisation constant: this process of constant adjustment is the cognitive process.

7　"Finally, I would like to add some sociological and ethical comments that I believe follow from understanding the autopoietic organisation of living systems. The essay on autopoiesis was supposed to have a second appendix on the social and ethical implications. This appendix, however, was never included because Francisco Varela and I never agreed on its content. I will now use the privilege I have in writing this introduction to present the notions I would have included in that appendix" (p. xxiii).

8　We notice that this problem we are discussing, the difference between a teleological model and one that accounts for progressive modifications, is the same as that posed in the evolutionary theory by the theory of "punctuated equilibrium".

9　The difference between signifier and signified might be unclear to the reader; therefore, we want to clarify it. A signifier is an object of the experience (also an actor) in the condition of taking on a meaning for another actor, namely to acquire a meaning in the cognitive domain. Not all the signifiers will actually acquire a meaning: those who will not will remain tacit, waiting for a systemic change in the socio-symbolic organisation, to emerge from the plasma. If a signifier acquires a meaning, it is therefore signified: the socio-symbolic system has included that signifiers among the objects/experiences that can be said.

10　From this point on, and even more so in the next chapter, we will use Latour's terminology almost interchangeably with that of Maturana and Varela. As we have repeatedly emphasised, our aim is to draw out the extreme consequences of ANT discourse by responding to the plasma problem. Terminologies and conceptual apparatuses often complement each other, as in the case of the sentence to which this note is attached. "Autopoietic networks of associations", for example, summarises both the associative functioning of actor networks described by Latour and the principle around which associations are organised, namely autopoiesis and its conceptual corollaries. Actors join networks and then become structurally coupled to them. The theoretical bridge linking the two theories is consideration, in the social, of the objects of the autopoietic organisation of the networks that we have defined as signifiers, i.e. actors and events that have the possibility of taking on one meaning or another depending on the autopoietic equilibrium of the network and its units/actors.

11　This implication opens up important scenarios for intersectional analysis (a point that goes beyond the purpose of the present work).

12　The reader should attend to the fact that the set of conceptual tools in this sentence, and in many others in this chapter, makes use of notions coming from Maturana, Badiou, and Latour simultaneously. We believe we have sufficiently shown in this work that these three systems of thought all point in the direction of understanding the functioning of social systems. As more than an exercise in style, we here offer the sentence in three versions faithful, respectively, to the language of each author, to show how they converge conceptually and how each thought illuminates areas that remain dark in the alternative formulations.

In Latour's language, we would read: (i)

> What is at stake, then, is to understand what process sometimes drives actors to suddenly reorganise themselves into a new network of associations, blatantly violating the continuity of intermediation and downloading plug-ins that completely upset the panorama that holds together the current association of networks in order to restructure themselves into a completely new one.

In Badiou's language, we would read: (ii)

> What is at stake, then, is to understand why the structure of Being is blatantly violated and subjects suddenly sign up to a new Truth, pointing to the void at the centre of the previous structure of Being and restructuring into a completely new one.

In Maturana and Varela's language, we would read: (iii)

> What is at stake, then, is to understand what process sometimes drives autopoietic units to blatantly violate the homeostatic and autopoietic principles, organising themselves into a completely new structural coupling that completely disrupts the previous autopoietic equilibrium.

What we wish to emphasise is that these three terminologies, which set out to capture the same thing, could not be brought into dialogue without the concept of the socio-symbolic system. This concept brings the dynamics described by the three approaches onto a semiotic level.

13 The disequilibrium is inexpressible because it cannot be signified. If it were signifiable, it would be part of the socio-symbolic organisation.

14 Matviyenko and Roof (2018, p. 53) also agree that Hayles' reading of Lacanian linguistics is somewhat reductive.

References

Bateson, G. (2000). *Steps to an ecology of mind: Collected essays in anthropology, psychiatry, evolution, and epistemology.* University of Chicago Press.

Bisconti, P., & Cesaroni, V. (Forhtcoming). *The pledge of the Ideo-logic. Anguish and psycho-epistemic between conspiracy theories and political ideologies.*

Guidano, V. (1992). *Il sé nel suo divenire. Verso una terapia cognitiva post-razionalista.* Bollati Boringhieri.

Guidano, V., Cutolo, G., & De Pascale, A. (2019). *La struttura narrativa dell'esperienza umana.* Franco Angeli.

Haraway, D. J. (2016). *Staying with the trouble.* Duke University Press.

Hayles, K. (1999). *How we became posthuman: Virtual bodies in cybernetics, literature, and informatics.* Chicago University Press.

Hayles, N. K. (1993). Virtual bodies and flickering signifiers. *October*, *66*, 69. https://doi.org/10.2307/778755

Holquist, M. (2003). *Dialogism: Bakhtin and his world.* Routledge.

Lacan, J. (2013). *The ethics of psychoanalysis 1959–1960: The seminars of Jacques Lacan.* Routledge.

Lacan, J. (2014). *Seminar X – Anguish* (J.-A. Miller, Ed.). Wiley and Blackwell.

Lacan, J., Miller, J.-A. E., & Fink, B. T. (2019). *Desire and its interpretation: The seminar of Jacques Lacan, Book VI.* Polity Press.

Latour, B. (1992). Where are the missing masses? Sociology of a few mundane artefacts. In W.E. Bijker & J. Law (Eds.), *Shaping technology/building society: Studies in sociotechnical change* (pp. 254–258). MIT Press.

Latour, B. (2005). *Reassembling the social: An introduction to actor-network-theory.* Oxford University Press.

Lettvin, J. Y., Maturana, H. R., McCulloch, W. S., & Pitts, W. H. (1959). What the frog's eye tells the frog's brain. *Proceedings of the IRE*, *47*(11), 1940–1951.

Maturana, H. R., & Varela, F. J. (2012). *Autopoiesis and cognition: The realization of the living* (Vol. 42). Springer Science & Business Media.

Matviyenko, S., & Roof, J. (2018). *Lacan and the posthuman.* Springer.

Mingers, J. (1991). The cognitive theories of Maturana and Varela. *Systems Practice*, *4*(4), 319–338. https://doi.org/10.1007/BF01062008

Simondon, G. (2020). *Individuation in light of notions of form and information.* University of Minnesota Press.

Volonté, P. (2017). Il contributo dell' Actor-Network Theory alla discussione sull' agency degli oggetti. *Politica & Società*, *1*, 31–58.

5 A machine politics

In the previous chapter we took a long break from social robots (SRs) and their social and political implications. It was indeed necessary to first understand why and how any technology could be able to modify the social and what mechanisms regulate these implications. In this chapter, we will address three political issues raised by social robotics in light of what has been stated above. First, we will explore how the machine can become a battleground between competing socio-symbolic systems, which inform different design strategies and different narratives about the purpose, risks, and benefits of SRs. The robot body, far from being politically neutral, is instead the field of a war of inscriptions, a proxy war that, in designing the robot, aims to retrospectively define the human. In the second section, we will discuss the political-eschatological function of SRs, taking as an example the analysis of Fadini et al. (2020), to understand how the narratives about robots constitute an aggregating *eschaton*. In the third section of this chapter, we will discuss a particular type of SR: the sexual robot. We will discuss the very idea of such an interactive machine and the implications for gender dynamics through the lens of Luce Irigaray's work. The sexual robot allows us to take a closer look at how interaction with a certain kind of machine design can impose consequences on the user's psychological organisation. This will lead us to Chapter 6 of this manuscript, analysing the intersubjective implications of SRs.

The field of a war of inscriptions

The robot represents a very particular kind of signifier within the symbolic economy of the societies that produce it. The most peculiar characteristic is certainly that of pre-existing the object it represents. The symbolic, mythological element of the robot in fact precedes the very existence of the object. In this sense, the signifier, as the union between the socio-symbolic element and the material support of signification, reverses the

DOI: 10.4324/9781003459798-5

typical process of the emergence and reorganisation of socio-symbolic structures. These structures usually reorganise the relations between their components following a telluric event of the real, which shakes the homeostasis of the system or increases its metastability. In this case, however, the relationship is reversed: the robot pre-exists in the form of the symbol in whose mould the ὕλη is fashioned. In this sense, the robot represents a chimaera to be realised. It is not the aim of this chapter to dig once again into the causes of the emergence of a signifier without environmental support. As discussed in the introduction to this manuscript, the ex nihilo creation of the human-like machine is a consequence of epistemic anguish, which need to separate the subject from the object and never blur the line. In the end of Chapter 4, we discussed how this creation ex nihilo can be a supporting element for socio-symbolic systems. In this chapter, however, we deal with the *implications* of such a reversal: the inverted relationship means that the symbol precedes the object. Therefore, this time it is the socio-symbolic equilibrium that shapes the outer environment when creating the object (the robot). Design procedures are therefore mediums that transform matter into the referent that best manifests the chimerical meaning of the symbol "robot". Design strategies shape the robot to best represent the material reference of a certain symbol chosen by the socio-symbolic system. An additional distinctive characteristic of the signifier-robot is that it is first and foremost a reference to the structure of the *anthropos* itself. In social robotics, anthropomorphism is one of the central elements in the design of the SRs, with the underlying belief that the greater the similarity to human beings, the greater the degree of sociality of which the robot is capable. We have already discussed how the anthropomorphic assumption is fundamentally an obstacle to the progress of social robotics as it does not take into account the multiplicity of interactional configurations that a context can assume. What we stress here, however, is that the anthropocentric teleology of robot design has an important normative aspect that applies not to the robot as such but to anthropomorphism as a category. The design of the anthropomorphic robot is a direct representation of the normative structures that a society expresses about the concepts of the human and the social. Against this backdrop, we agree with (Devlin, 2016) that "a machine is a blank slate that offers us the chance to reframe our ideas". In his article, Keith Devlin means this phrase in a positive sense, but we emphasise its political relevance. The operation of anthropomorphising the robot has the triple effect of producing a normative structure that (i) defines what it means to be "human" and what it means to have a "social relationship"; (ii) carries out this operation in an alienated and surreptitious fashion on a supposedly neutral ground, namely robot design; and (iii) reveals how the robot is an extremely plastic signifier. Its design is a tool for an ideological "proxy war".

Let's address these topics in order. The designer of the robot or of its interaction, when formulating the hypothesis that a certain phenomenal element (material or behavioural) can increase or decrease the robot's anthropomorphism, is first of all performing a normative operation defining *what features are anthropomorphic*. This applies not only strictly to a given purportedly anthropomorphic feature *per se* but also to the whole universe of such features, including *trustworthiness, competence,* and *likeability*. This operation has the retroactive effect – for example, in the choice of features for measuring interaction – of defining those elements that define the human as such. It means establishing categories for measuring the difference between human and non-human. Features associated with the latter are never categorised in the proper sense but only excluded from the anthropomorphic group. At the same time, therefore, two normative aspects pop up: that of design, i.e. how the robot is made and how it behaves; and that of measurement, i.e. how to evaluate the extent of the robot's compliance with the set of anthropomorphic categories. These kinds of considerations aim to show the normative nature of the design operation and its retroactive implications in defining the category of *anthropos* itself. This normative operation on what is human is carried out surreptitiously, in a dislocated position, namely on the robot. This implies that the direct effects on the concept of "human being" are more difficult to trace. The normative design has very concrete implications: one example discussed in the literature is the relationship between gender, robotics, and the social function of the robot. The differentiation into robots with male and female features often has a direct co-implication with the social function for which the robot was built (Bisconti & Perugia, 2021). Robots that perform care functions are typically associated with female traits (Robertson, 2010), while robots that have managing functions or should be perceived as authoritative have masculine traits (Bernotat et al., 2017). As social awareness of gender issues has intensified, this topic has finally reached the community of SR designers, who are now discussing how to avoid reinforcing sexist and exclusionary social norms. The fact that, in socio-technical systems, objects and their design are charged with socio-cultural biases is nothing new: the scientific community studying forms of artificial intelligence, such as machine learning algorithms, now vigorously discusses the replication of exclusionary norms in algorithm datasets (Wachter-Boettcher, 2017). What makes SRs structurally different is that they resemble humans, acting as dislocated objects normatively defining the human. Being social actors places SRs in a limbo of particular complexity with respect to other non-human actors in socio-technical systems. The supposedly neutral terrain of the design of the SR ultimately shows the co-implications between descriptive and prescriptive practices with respect to the categories relevant to defining the human. This difference is by no means irrelevant to the process of socio-symbolic mediation that the SR can provoke. Ultimately, we should reject the precedence of the structure of the symbol over the material referent. In fact, we recognise that the material referent of the signifier "social robot" is by no means only the robot but includes above all the human being. For this reason, from a political point of view, the robot represents a field of normative descriptions, with direct reference to the human as the field of prescriptive operations. As in the case of a proxy war, the field of definition of the human is displaced. The risk we face is that of all dislocated and surreptitious constructions: the process of alienation is constituted in a way that is no longer traceable and is itself elided by public discourse.

Technologies and eschatologies

In the previous section we addressed the politics of SRs from the point of view of the signifier–symbol relationship and its surreptitious reference to the human. In this section we discuss the eschatological dimension of robotics in the society's narratives. We already addressed above the close relationship that exists between eschatology and homeostasis, which implies the use of certain signifiers as Truths/Events, which act as symbolic collectors of numerous units and socio-symbolic associations. We have also suggested that signifiers grounded "in absence" are ideal candidates to become eschatological collectors of networks as they lack material support and are fit for multiple autopoietic organisations.

In this section, we compare typical narratives about robots and AI in order to understand how these narratives work. We draw on Fadini et al.'s work (2020), which discusses the supposed advent of so-called *Superintelligence*. This concept, first mentioned in Bostrom (2017), supports the view that in the near future, artificial intelligence algorithms, especially those based on machine learning, could "take over" and endanger the very existence of humanity. A not-too-dissimilar argument is made about robotics. According to a famous study by Frey and Osborne (2017), in the next 30 years, robotic automation will replace most jobs. Interestingly, these eschatological narratives are always presented in a millenarian fashion, whether positive or negative: the end of the world is near. Sometimes, that end of the world is depicted as desirable because it ends *this* world and allows the beginning of a better *new one*: "Thanks to robots, human beings will finally be free from the burden of work"; "Artificial intelligence will lead us to a new, transhuman era of existence". At other times, it is described in a negative sense: "Automation will cause wars and destruction because people will no longer be earning a living"; "AI will destroy the world because it will rebel against its creators". In any case, these eschatological and millenarian symbols are structured as events that bring history to an end, at least as we know it: they either annihilate humanity completely or free it from its problems once and for all. These narratives then serve as a Trojan horse to convey content of dubious ethicality and desirability, as Henry (Fadini et al., 2020) notes:

> the explicit eugenic approach of the hoped-for use of genetic engineering, the only truly robust alternative, through the enhancement of humanity's general and collective intelligence, to the emergence of a legibus soluta AI, of a global cybernetic Leviathan, called singleton, is not lost on us. [...] In Bostrom's approach, the moral and religious reluctance to accept eugenics in order to enhance the intellectual faculties of the human race appears to be an insurmountable constraint, and thus a structural bias, which, even if one wanted to, one could not ignore. Reading between the lines, one deduces that, in the best of all possible worlds, only the elimination of such qualms would put our species in a condition of security that would be impervious to the emergence of a superhuman, sovereign and non-benevolent AI.
>
> (Translation is mine)

Thus, eschatologies have as their immediate effect the narrative legitimisation of real actions (eugenics) against imaginary threats (the *AI singularity*). With regard to robotics, the fear of full automation leads to conservative and Luddite drifts, which have the effect of justifying the employment of human labour in wearisome tasks. These are both ways of justifying the status quo, based on the phantasmatic support of the subsistence of an eschatological signifier. Social robotics, meanwhile, becomes the field of a struggle between narratives about the future of sociality: SRs will either completely replace human relations, definitively disintegrating sociality, or they will become the salvation of the elderly, definitively curing their loneliness. Both these standpoints take no account whatsoever of the SRs' current technical status. The most advanced SRs still cannot easily move in an environment without facing insurmountable obstacles, nor can they carry on a conversation that does not follow a rigid script.

Yet both these portrayals take the form of a fetishistic narrative device that polarises public discourses on technology. We believe that there are two reasons for the effectiveness of these apocalyptic narratives. The first is clearly stated by Tomasetta when he argues that

> a crucial element in clarifying the reasons for the appeal of apocalypses lies in the fact that these myths offer those who follow them a sense of absolute centrality. [...] This is a sort of chronological equivalent of ethnocentrism, a belief that is both reassuring and exciting, which removes the threat of insignificance from our presence in time.

The second reason is summed up by Henry (Fadini et al., 2020) when she argues that *Superintelligence* underlines:

> the tension between anthropocentrism and the limits to our power over non-human entities, the ecosystem, the cosmos; the conflict between an epistemic deference to the cognitive perfection of a wiser AI – because it is free from base instincts and base motivations – and the terror of losing control over it precisely because of the non-anthropomorphic features of its essential configuration.
>
> (Translation is mine)

Here again we see a *leitmotiv* of our discussion: the eschatological narrative is constituted as the phantasmatic support of a contradiction. On the one hand, humans consider themselves omnipotent creators ex nihilo so as to separate themselves from the material and the *transeunte*. On the other hand, we find an unexpected effect of this movement: the line dividing subject and object is blurred. In *Superintelligence* the content of the phantasmatic support is the cognitive capacities of the human being; in social robotics it is instead the entry of the object into the realm of the social. According to Henry's reasoning, the problem lies not in relying on a phantasmatic support, which is also common to *science fiction*, but in the cloaking of this phantasmatic support in scientificity. Bostrom's and Frey and Osborne's (2017) dystopias are presented as a prediction, not a speculative narrative, of the years to come. This calls for direct political consequences: since it is (ostensibly) a reliable prediction, *policy* measures (such as eugenics) must *now be put in place* to avoid the apocalypse.

Even if these narratives were constituted as positive millenarianisms, the result would not be qualitatively different. Accelerationism, typical of prophets of full automation, is a powerful device for defusing or normalising today's political contradictions, such as that of inequality. The promise of full automation is a world without conflict, full of abundance, in which inequalities are eclipsed by a new proletarian army consisting of robots. Tireless and without union demands, robots become the phantasmatic support of a society that heals its internal contradictions with the intervention of a *deus ex machina*. In such visions, SRs definitively cure social contradictions, heal the increasing loneliness highlighted in *Alone Together* (Turkle, 2017), or perhaps bring the human to solipsism, as in the cartoon *Wallie*. The promise in any case is: everything will be completely different from today. This mechanism narratively cancels the existing contradictions and defuses their conflictual content, postponing the resolution to a day "after history".

The function of catastrophic pseudoscientific predictions within the economy of the socio-symbolic is now clear: they support a void on the side of the signifier. We claimed that the socio-symbolic is constituted by the process of signification of material, iletic data within the medium of autopoietic structural coupling. When this becomes unbalanced due to a change in the real, the metastability of the system increases, and the ideal homeostatic equilibrium point is distanced. In order not to change its organisation, the socio-symbolic system enrols signifiers without reference in the system's environment. These support the socio-symbolic equilibrium, covering a contradiction between the system and its environment, so that the current organisation can be preserved. We can now propose a definition of the signifiers constituted in the absence of a material referent: these are the phantasmatic supports of the socio-symbolic system.

Sex robots: the phantasmatic support of masculinism

The uses of SRs can be diverse, given the interactional capacity that some of these machines currently achieve. For the time being, they are mainly used to deal with the relational deficiencies that some subjects experience – in most cases weak segments of the population, within the limits of SRs' effectiveness. Most attention is certainly being devoted to assistive robotics for the elderly (Heerink et al., 2010; Kachouie et al., 2014; Turkle et al., 2006); another flourishing field of research focuses on interaction with children on the autistic spectrum (Coeckelbergh et al., 2016; Matarić & Scassellati, 2016; Peca et al., 2016). In recent years, the topic of sexual robotics has been gaining attention in academia and in the realm of public opinion.

Sexual robots (SexRs), usually with a female appearance, aim to reproduce human sexual interaction as believably as possible (Sullins, 2012). They are designed to replicate not only the physical aspects of a sexual relationship but also the relational and emotional ones. To date, there are few examples of SexRs on the market, and they are still in an early stage of development (Danaher, 2017). These robots can interact physically and verbally with users, using a predefined set of behaviours (Bendel, 2017). The starting point of the academic discussion on sexual robotics may be located in Levy's book *Love and Sex with Robots* (Levy, 2007), where the author analyses the nature of human-robot sexual interactions in comparison with human–human sexual interactions. His conclusions reflect enthusiasm about the introduction of this technology. Subsequently, interest in sexual robotics has spread in the scientific literature. The main concerns relate to the possibility of raping a SexR (Strikwerda, 2015), the commercialisation of child SexRs (Maras & Shapiro, 2017), and their moral status. A highly relevant topic of discussion is the social impact of this new technology: SexRs could change the social understanding of sexual consent and the social representation of women.

Indeed, the majority of sex robots marketed today are female in appearance (Cox-George & Bewley, 2018), and this trend is likely to persist for a long time. According to many scholars, the possibility of freely using a female robotic body and performing degrading or violent practices on it could change the social representation of women (Gutiu, 2016). In general, the discussion is divided between supporters and detractors ofSexRs. The former argue that rape and violence will decrease (Devlin, 2016), users' sexual satisfaction will increase (Levy & Loebner, 2007), and SexRs will contribute to the reduction of prostitution (Levy & Loebner, 2007). The latter argue that SexRs will contribute to increasing violence (Sparrow, 2017) and will reproduce existing gender inequalities (Cox-George & Bewley, 2018; Gutiu, 2016).

Elsewhere, we have discussed the psychological implications of SexRs (Bisconti Lucidi & Piermattei, 2020; Bisconti, 2021). In this section, we deal with the political implications of the existence of such machines. This analysis will focus on two questions. The first concerns the structure of the sexual robot as a symbol. We investigate it through the thought of Irigaray, relating SexRs to the phallologocentric exclusion of the thinkability of women. We will conclude that the sexual robot is a phantasmatic support of masculinism. The second question concerns the effects of such a support for the socio-relational equilibrium. This last analysis will lead us to the theme of the next section: the challenge that the relational robot poses to intersubjectivity.

Of great relevance for the analysis of SexRs, symbolic function is the thought of Irigaray, especially her most important contribution to the "feminism of difference", namely the extended essay *Speculum of the Other Woman* (Irigaray, 1985). The theme around which this work is articulated is the concept of phallologocentrism. This term indicates three intertwined points: that the phallus and the logos are in some respects one, that they belong to a male symbolic production, and that they are at the centre of ideological discourse. The identification between phallus and logos is a Lacanian theoretical contribution that Irigaray takes up in a critical form. Phallus and logos are mirrors of each other. They are an ideological discursive production wherein the first is a semantics, and the second a syntax of female exclusion from thinkability and from the possibility of autonomous signification. The first provides the content of the law of patriarchy: the law of the phallus, which in Leví-Strauss means woman as an object of exchange for clans (Lévi-Strauss, 2003). The law of the phallus constitutes that metaphysical and symbolic horizon that makes the male subject an objectified subject thinking of himself. Against this backdrop, Irigaray's critique concerns philosophical thought *tout court*, identified as the maximal manifestation of the universalistic self-production of phallologocentric thought. Like a celibate machine, the masculine dynamically self-produces in order to cover its self-referentiality. The feminine is a product of this thought: the inessential opposite, a locus of dialectical-negative development of the masculine. The phallus therefore produces a "grammar of exclusion": the logos, which acts especially in the theoretical realm of philosophy. The Other of the masculine subject, the feminine, is excluded from thinkability. The feminine turns out to be a *speculum* in phallologocentric thought, an instrument of narcissistic reflection of the masculine. This Sameness, the masculine, produces an Otherness, the feminine, that does not belong to women. Females are described as stateless subjects in the symbolic horizon of males. They suffer a double deception: alienation from their essence and being forced to accept the "feminine" produced by phallologocentric thought. The masculine therefore acts with the double movement of abstracting and then universalising, as in the definition of "man", which is simultaneously all human beings and only masculine ones. Most of the engineers who design robots are male, and there is an ongoing debate on the inclusion of women in the design of technology.[1] So who is producing the designs of sexual robots, and in the image of what? The feeling, looking at the first marketable prototypes, such as Roxxxy, is that sexual robots are an obscene and grotesque version of the representation of masculine desire. In short, sexual robots are too obscene[2] to play the role of a mere substitute for a human sexual partner. We claim, contrary to most of the current literature, that sexual robots are not designed to replace the sexual relationship between two individuals and do not represent a simulacrum of the sexual partner. We claim that SexRs substitute something different from a human partner.

Before going further, what is the meaning of "obscene" in this context? Partly, it is the Greek sense of "outside the scene":[3] the scene outside which SexRs are placed is the dynamic self-production of the phallus, which continually produces the feminine as an otherness, only to bring it back into the phallologocentric structure as a negative in which to self-produce. Against this backdrop, the alleged "naturalness" of sexual difference is itself a form of subrection (and therefore of masking) of this self-production. The phallologocentric production of the feminine is a sustainable fiction only up to the moment in which it *conceals* the process of alienated production. Sexual robots, as a symbol, fail precisely here. They are an unmaskable fantasy concerning not the Other as such but the male subjects' structure of desire. This structure elides the Other insofar as the sexual robot has the fundamental characteristics of total acceptance and passivity with respect to the will of the user, his universe of desire. What it represents, as a simulacrum, is not the feminine but the omnipotence of the male subject. In short, in the obscene representation, we find the images of the phallus and of self-referral. This last process, which Irigaray highlights clearly in the universalisation of phallologocentric philosophical thought, can be found in a less complex form in the design procedures of sexual robots. In fact, the self-production of the normative sexual structures of society is visible in the design of anthropomorphic "female" robots such as gynoid maids, assistants, and receptionists. The design of sexual robots has more radical normative consequences. For it no longer conveys a normative representation of women but directly conveys the desire structure of phallologocentrism, now openly not referred to an alterity. Sexual robots in fact enable the structure of desire to unfold in the absence of moral qualms, based for example on the principle that damage does not apply to human–robot interaction (HRI). The result is a representation of the feminine, in the sexual robot, that is completely at the disposal of the user, an omnipotent relational dynamic. Against this backdrop, the sexual robot is the phantasmatic support of masculinism: it is a *simulacrum* that enables the fantasy that the victim is not only docile and available but even positively responsive in the relationship. An example of this is that one of the "personalities" that Roxxxy can play is a "submissive" one that willingly and passively accepts any kind of sexual practice performed on her body. The robot is thus structured as a phantasmatic support of the fantasy of omnipotence, which concerns not the representation of a real woman but her objectification as the object of another's desire.

The issue that arises at this point concerns two deeply interconnected questions, which will be the subject of the next of this manuscript. The first concerns the status of the relationship between the user and the robot, be it sexual or simply social. From what we have discussed, it is now clear that the robot is first and foremost a signifier that *supports* certain discursive practices of the social. The second fundamental issue is that the SR is a *support* that speaks, and this constitutes a huge difference in its ability to mediate the socio-symbolic system in comparison to other objects. The social condition of objects, namely their function in networks, does not usually allow them to "act" on an equal footing with humans. Objects can indeed provoke modifications to the system, especially on the side of the material support of the signifier, but they are not usually found on the side of symbolic-semantic configuration as speaking actors. In contrast, SRs can say something about themselves. They are interactive and relational on a material, proxemic, and even linguistic level. What kind of relationality does a talking object perform? This will be the topic of the next section's discussion. We will enquire whether intersubjectivity is still an effective paradigm for interpretation or whether a particular form of alienation is a better one. From this discussion a further question will immediately follow: how are individuals' relational settings modified as an effect of human–robot relations? A relational object, when producing a new form of sociality, does not have neutral consequences for all the other relational configurations experienced by the subject. This is a basic assumption following from the notion of socio-technical systems. Therefore, we will conclude by discussing the extent to which human–robot relations can modify human–human relations, and how they can do so.

These are the topics of the next chapter of this manuscript.

Notes

1 www.fastcompany.com/1665597/how-women-are-leading-the-effort-to-make-robots-more-humane
2 www.independent.co.uk/life-style/sex-robots-frigid-settings-rape-simulation-men-sexual-assault-a7847296.html
3 This etymology is actually subject to considerable debate, but we encourage the reader to consider its metaphorical value ahead of its etymological accuracy.

References

Bendel, O. (2017). Sex robots from the perspective of machine ethics. *Lecture Notes in Computer Science (Including Subseries Lecture Notes in Artificial Intelligence and Lecture Notes in Bioinformatics)*, 10237 LNAI, 17–26. https://doi.org/10.1007/978-3-319-57738-8_2

Bernotat, J., Eyssel, F., & Sachse, J. (2017). Shape it – The influence of robot body shape on gender perception in robots. In A. Kheddar, E. Yoshida, S. S. Ge, K. Suzuki, J.-J. Cabibihan, F. Eyssel, & H. He (Eds.), *Social robotics* (pp. 75–84). Springer International Publishing.

Bisconti Lucidi, P., & Piermattei, S. (2020). Sexual robots: The social-relational approach and the concept of subjective reference. In M. Kurosu (Ed.), *Human-computer interaction design and user experience: Vol. 12182 LNCS* (LNCS, pp. 549–559). Springer. https://doi.org/10.1007/978-3-030-49062-1_37

Bisconti, P. (2021). Will sexual robots modify human relationships? A psychological approach to reframe the symbolic argument. *Advanced Robotics*, 35(9), 561–571. https://doi.org/10.1080/01691864.2021.1886167

Bisconti, P., & Perugia, G. (2021). How do we gender robots? Inquiring the relationship between perceptual cues and context of use. *Proceedings of GenR 2021 Workshop on Gendering Robots: Ongoing (Re) Configurations of Gender in Robotics*, 4, 1–6.

Bostrom, N. (2017). *Superintelligence*. Dunod.

Coeckelbergh, M., Pop, C., Simut, R., Peca, A., Pintea, S., David, D., & Vanderborght, B. (2016). A survey of expectations about the role of robots in robot-assisted therapy for children with ASD: Ethical acceptability, trust, sociability, appearance, and attachment. *Science and Engineering Ethics*, 22(1), 47–65. https://doi.org/10.1007/s11948-015-9649-x

Cox-George, C., & Bewley, S. (2018). I, sex robot: The health implications of the sex robot industry. *BMJ Sexual & Reproductive Health*, 44(3), 161–164. https://doi.org/10.1136/bmjsrh-2017-200012

Danaher, J. (2017). The symbolic-consequences argument in the sex robot debate. In J. Danaher & N. McArthur (Eds.), *Robot sex: Social and ethical implications* (pp. 170–196). MIT Press.

Devlin, K. (2016). In defence of sex machines. *The Conversation*, 2015–2017. http://research.gold.ac.uk/17310/1/In defence of sex machines- why trying to ban sex robots is wrong.pdf

Fadini, U., Henry, B., & Tomasetta, A. (2020). Superintelligenza. Tendenze, pericoli, strategie di Nick Bostrom. *Iride, 2*. https://doi.org/https://doi.org/10.1414/98550

Frey, C. B., & Osborne, M. A. (2017). The future of employment: How susceptible are jobs to computerisation? *Technological Forecasting and Social Change, 114*, 254–280. https://doi.org/10.1016/j.techfore.2016.08.019

Gutiu, S. M. (2016). The roboticization of consent. In *Robot law* (pp. 186–212). Edward Elgar Publishing. https://doi.org/10.4337/9781783476732.00016

Heerink, M., Kröse, B., Evers, V., & Wielinga, B. (2010). Assessing acceptance of assistive social agent technology by older adults: The Almere model. *International Journal of Social Robotics, 2*(4), 361–375. https://doi.org/10.1007/s12369-010-0068-5

Irigaray, L. (1985). *Speculum of the other woman*. Cornell University Press.

Kachouie, R., Sedighadeli, S., Khosla, R., & Chu, M.-T. (2014). Socially assistive robots in elderly care: A mixed-method systematic literature review. *International Journal of Human–Computer Interaction, 30*(5), 369–393. https://doi.org/10.1080/10447318.2013.873278

Lévi-Strauss, C. (2003). *Le strutture elementari della parentela*. Feltrinelli Editore.

Levy, D. (2007). *Love and sex with robots*. HarperCollins.

Levy, D., & Loebner, H. (2007). Robot prostitutes as alternatives to human sex workers. *IEEE International Conference on Robotics and Automation*.

Maras, M.-H., & Shapiro, L. R. (2017). Child sex dolls and robots: More than just an Uncanny Valley. *Journal of Internet Law, 21*(6), 3–21.

Matarić, M. J., & Scassellati, B. (2016). Socially assistive robotics. *Springer Handbook of Robotics, 18*(1), 1973–1993. https://doi.org/10.1007/978-3-319-32552-1_73

Peca, A., Coeckelbergh, M., Simut, R., Costescu, C., Pintea, S., David, D., & Vanderborght, B. (2016). Robot Enhanced therapy for children with autism disorders: Measuring ethical acceptability. *IEEE Technology and Society Magazine, 35*(2), 54–66. https://doi.org/10.1109/MTS.2016.2554701

Robertson, J. (2010). Gendering humanoid robots: Robo-sexism in Japan. *Body and Society, 16*(2), 1–36. https://doi.org/10.1177/1357034X10364767

Sparrow, R. (2017). Robots, rape, and representation. *International Journal of Social Robotics, 9*(4), 465–477. https://doi.org/10.1007/s12369-017-0413-z

Strikwerda, L. (2015). Present and future instances of virtual rape in light of three categories of legal philosophical theories on rape. *Philosophy & Technology, 28*(4), 491–510. https://doi.org/10.1007/s13347-014-0167-6

Sullins, J. P. (2012). Robots, love, and sex: The ethics of building a love machine. *IEEE Transactions on Affective Computing, 3*(4), 398–409.

Turkle, S. (2017). *Alone together: Why we expect more from technology and less from each other*. Hachette UK.

Turkle, S., Taggart, W., Kidd, C. D., & Dasté, O. (2006). Relational artifacts with children and elders: The complexities of cybercompanionship. *Connection Science, 18*(4), 347–361. https://doi.org/10.1080/09540090600868912

Wachter-Boettcher, S. (2017). *Technically wrong: Sexist apps, biased algorithms, and other threats of toxic tech*. W. W. Norton.

6 The intersubjective dimension of human–robot interactions

This is certainly not the first essay to question the possible implications of social robotics for users' lives. There are many points of view, angles of analysis, and answers to this question, as there are many articles discussing the subject. From the literature that discusses the future use of social robots (SRs), one can extrapolate five major criticisms or concerns expressed by scholars (Sharkey & Sharkey, 2012). We will summarise and comment on these main arguments. We will mainly discuss the so-called *deception objection*, which will then be reformulated within a different understanding of intersubjective relations. It is necessary, in commenting on these five most critical points, to distinguish between undesirable consequences deriving specifically from SRs as such and implications that are common to many other technologies. In short, we believe that it is important, when discussing the effects of SRs, to carefully differentiate between the changes that robotics causes *suo motu* and those that were already present as general trends in socio-technical systems, which robotics might help to manifest. This distinction makes no less relevant the discussion of how to minimise the impact that robots might have on the social, cultural, and political processes already in place. The resolution of social implications is a key aspect of a successful inclusion of robots in the everyday lives of millions of people, even if these consequences are not specific to robotics. However, the scope of this work is limited to the socio-systemic and psychological peculiarities of interacting technologies. Therefore, we will only briefly mention the other issues. The five main concerns reported by Sharkey and Sharkey (2012) are:

1 The potential reduction of human contact, i.e. the likely progressive loss of interaction with other humans due to human–robot interactions (HRIs).
2 A loss of privacy due to constant monitoring.

DOI: 10.4324/9781003459798-6

3 Loss of personal freedom due to the physical control of the robot over the user.
4 *Deception*: robots with the ability to simulate mental states, to interact verbally (and non-verbally) with humans, to understand and respond to their emotional states, raise the question of whether it is legitimate to deceive humans about the simulated nature of intersubjective robotic interactions.
5 The problem of control over robots: under what circumstances should users be able to control the robots they are interacting with?

This last problem is particularly urgent with regard to *assistive robotics* for the elderly and children. There is certainly a need for a way for those under a robot's care to deactivate it as mechanical failures or software errors can jeopardise the person's life. On what occasions should elders be allowed to "switch off" a SR? If they were always allowed to do so, the coercive component that may be necessary in the treatment of psychologically unstable individuals would be removed. Even without considering a robot's hardware or software errors, some actions generally denied to the elderly person, such as leaving the house in a hurry, may be necessary in extraordinary cases such as that of a gas leak in the house. Obviously, a robot can hardly understand with the same precision as a human being all the contextual aspects that need to be considered before allowing someone to break a rule or denying that ability.

It is easily understandable how most of these concerns are not raised specifically by SRs. Privacy and security issues are common to all technologies. On the other hand, the issue of user deception about the artificial nature of SRs raises an entirely new question within the landscape of unintended implications of technological innovation. Never before has it been necessary to deal with objects that are able to simulate human interaction faithfully. Moreover, while the issue of diminishing human contact is certainly not unique to social robotics (Turkle, 2017), there is still a strong argument that SRs may worsen the problem far beyond the current interactive technologies. This section will therefore address these two issues, discussing and reformulating the validity of the deception objection and analysing the intersubjective structure of HRIs.

To start with the first point, the so-called *deception objection* has been raised numerous times in the philosophical and ethical literature on social robotics (Coeckelbergh, 2010, 2018; Sætra, 2020; Sharkey & Sharkey, 2010). The most classic formulation is certainly that of Sparrow and Sparrow (2006):

What most of us want out of life is to be loved and cared for, and to have friends and companions, not just to believe that we are loved and cared for, and to believe that we have friends and companions, when in fact these beliefs are false.

(p. 115)

We have already raised our concerns about the *deception objection in another work* (Bisconti Lucidi & Nardi, 2018). In the next few pages, we will briefly run through them and then formulate an initial version of the self-deception hypothesis, which will guide the rest of this chapter.

Rethinking the deception objection

Discussion of the possibility of a robot deceiving a human about its artificial nature is consubstantial with the whole imaginary, both scientific and narrative, about humanoid machines. In the fashion of a challenge to be overcome, the Turing Test measures the ability of a machine to be unrecognisable as such in an interaction with a human being. As far as social robotics is concerned, the academic concerns of "robotic deception" began just when social robotics was making decisive advances in the development of humanoid robots, creating the first robots capable of interacting socially with a human. Although the first papers dealing with the problem of the deception objection date back to the early years of the millennium, the literature on the subject has increased exponentially since the publication of Sparrow and Sparrow (2006).

We begin with a few comments on how the discussion of the deception objection has been set up from the beginning and how it is still taking place to this day. The academic conversation on the subject has mainly involved experts from the social sciences and humanities, especially philosophers and lawyers. The total lack of interdisciplinarity is particularly evident. The humanities community has constantly warned of the possibility of a robot's deceiving a human and the consequent deceitfulness of the emotional and sympathetic aspects of interaction. These warnings seem to take no account of the current stage of development of interactive robots. Not only are SRs not currently able to deceive a human about their artificial nature; there is also nothing to suggest that this problem is likely to arise in the near future. The current interactive capability of a NAO or a Pepper, two of the world's most popular SRs, is extremely rudimentary in terms of both physical interaction with the environment and their verbal and non-verbal interactive capabilities. Moreover, not a single actual case of deception by a SR is reported in the literature: no user has ever been convinced that a robot had a human or living nature. Supporters of the deception objection claim that such a deception could occur in socially weaker population groups susceptible to anthropomorphising, especially children and the elderly. In the literature on assistive robotics for elders, however, we cannot find any reference to actual deception but only to cases that, for now, we will define as self-deception.[1] A particularly striking example of this can be found in Turkle's pioneering study (Turkle et al., 2006) "Relational artifacts with children and elders: The complexities of cybercompanionship", in which the elderly Andy remains fully aware of the artificial nature of the robot (a MyRealDoll) despite projecting his ex-wife onto it. The children taking part in the experiment also engage in various degrees of anthropomorphism, but in no case do they consider the robot to be a human being.

Notwithstanding the unlikeliness of deception, this term has nonetheless unfortunately attracted an important portion of the discussion of the social, relational, and psychological implications of SRs. This unjustified concern can be brought back into the conceptual framework of the previous chapters: the anxiety provoked by the entry of objects into the social sphere and that of technological millenarianism. The topic of deception is in fact an excellent example of that mechanism. What the deception objection suggests is that there is, on the one hand, a true, authentic, and "pure" relationality, represented by the relationship between human beings, while on the other hand, an impostor robot could mimic and parasitically appropriate the forms of human relationality in the absence of their substance. The theme of a parasitic and invisible enemy, which invades the community by mimicking its behaviour and slowly colonises it, is a common theme in literature and fiction. Two sides of this narrative are interesting to analyse. The first is certainly the troubling aspect generated by the narrative of the caretaker robot, which enacts emotional behaviours without actually feeling the emotions. This directly recalls the notion of the uncanny in Jentsch (1997): the doubt whether something is dead or alive. The interesting point is not simply that the SR can present as uncanny, a fact that has been obvious since the famous paper of Mori (1970), but rather that this topic has become the dominant narrative. Like a socialised Capgras syndrome,[2] this narrative has social robotics populating the world with mechanical doppelgangers, polluting the (supposed) purity of human intersubjectivity. Some scholars (Sharkey & Sharkey, 2021) have already pointed out the difficulty of supporting the position of considering the deception itself and not its consequences. Moreover, the deception objection approach also operates a surreptitious division: relationships with machines are said to be "false", while those between humans are said to be "true" and authentic (Bisconti Lucidi & Nardi, 2018). We raise two criticisms of this view:

1 It is based on an implicit assumption that sublimates intersubjective relations between human beings, considering these relations as "pure" and devoid of any degree of deception.

The behaviour of subjects, including relational behaviour, is actually driven by an extremely vast number of motives that go beyond pure and disinterested relationality. The reasons for the way a human relates may variously be psychological, derived from social norms or dictated by self-interest. Thus, there is no such thing as pure, authentic, and disinterested human relationality. This should lead us to turn the deception *objection* towards human beings, arguing that any interaction carried out on the basis of stimuli that are not morally pure is deceptive. Indeed, the objection presupposes a surreptitious and sublimated subjectivity, emptied of all the unconscious, cultural, and normative aspects that determine a subject's behaviour in reality. And this argument is clearly untenable.

2 The objection presumes, without support, that the ethicality of relational behaviour should be measured by the subject's intention rather than by the behaviour's effect.

The intentional structure of the subject does not necessarily inform the consequences of the action. An SR may be benefiting to a user, while being deceiving. SRs for the elderly are certainly "tricking" the user when they respond positively to an emotional stimulus (e.g. responding affectionately to a caress, like Paro), yet these subjects may benefit from the interaction. Moreover, they might enjoy the "emotional" interaction even knowing that is artificial, as some papers show (Turkle et al., 2006).

Therefore, on the one hand, the deception objection focuses exclusively on the intentional structure of relational actions; on the other, it surreptitiously pits human relationality, seen as pure and disinterested, against human–robot relationality, seen as deceptive and malignant. These two characteristics of the deception objection, among others, distract and prevent us from investigating the most urgent implications of SRs: the sociotechnical and relational effects that current SRs – or those presumably available in the next few years – may cause.

Finally, to repeat, we claim that the deception objection is an eschatological narrative in the manner of Bostrom's *Superintelligence*, locking the discussion of SRs into implications that SRs do not currently raise and perhaps will not raise for a very long time to come. This process takes the focus away from the problems that HRIs raise today, since these are far less fascinating than the rhetoric of deception.

The current formulation of the deception objection is to be rejected, both for its moralistic point of view and for reasons of inexact conceptualisation of the problem. However, the issue that concerns social scientists is the effects that social robotics could have on human social and relational structures.

While the possibility of a SR's deceiving its user is remote, the likelihood that it will alter human relationality is to be taken seriously. Indeed, Turkle's (et al., 2006) elderly Andy gradually isolates himself from the community and begins to spend more of his time with his MyRealDoll, lost in his thoughts of reconciliation with his wife. In the words of Shannon Vallor, what SRs do is "shaping human habits, skills, and traits of character for the better, or for worse" (Vallor, 2016). A particularly striking example in the discussion of the social and symbolic consequences of SRs is sexual robotics, which we have already discussed in part in the previous section. In the discussion of sexual robotics, the argument of the so-called symbolic shift is formalised (Danaher, 2017). This argument supports the possibility that negative symbols or practices stimulated by sexual robotics may be transferred beyond those interactions and shape social behaviours. Gutiu (2016) argues, for example, that sexual robots, by reproducing patriarchal behaviours and symbologies, will normalise these behaviours in society. Starting from a virtue ethics approach, Peeters and Haselager (2019) put forward the same claim: to perform certain behaviours within a certain symbolic framework, for example a patriarchal one, has the effect of normalising the symbolic content outside of that context. As we have already discussed elsewhere (Bisconti, 2021), these approaches do not question the criteria for symbolic transfer and are not fully convincing.

The mechanism at the root of symbolic transfer, formulated mainly in discussions of sexual robotics, seems to respond to a simple behavioural automatism: if a subject performs an action or sees it performed a certain number of times, then it becomes normalised and therefore transferred to the behavioural patterns of individuals and the symbolic structure of society. In fact, this presumption of the automaticity of symbolic shifts is not *explicitly* supported by the authors cited above. At no time do they give an explanation of the processes of this transfer or, above all, of their non-occurrence in certain cases. Our view is that when describing the principles of the occurrence of a phenomenon, it is first necessary to understand the principles of its non-occurrence. Accordingly, in analysing socio-symbolic crises above, we focused first on why they often fail to occur. Only in this way, by focusing on the principle of the non-occurrence of a phenomenon, can we be confident that our theory is well grounded. An excellent example of a non-occurring symbolic transfer is theatre.

Suppose an actor impersonates a sadistic murderer every night on the stage for months. This actor nightly recites immoral discourses and simulates heinous murders. Although not all actors follow Stanislavsky's approach, a certain degree of impersonation is certainly required to play a role. If we gave full credit to the theory of behavioural automatism, we would expect such actors, at some point, to transfer the brutality of their character offstage into their private lives. Not only does this not happen, but we believe it is also in strong collision with the notion of catharsis. Catharsis gives an opposite account of the theatrical phenomenon as a context in which certain desires can be acted out, discharging the corresponding drives precisely in order *not to* commit the same actions in reality. In fact, there is a well-known case in which catastrophic predictions of the transfer of behavioural patterns have proved false: violent video games. Drummond et al. (2018) show how discussion of the alleged increase in aggression due to violent video games has failed to demonstrate a link between the two phenomena. Rather, some studies even suggest an inverse correlation (Markey et al., 2015). In short, behavioural automatism does not seem to be a good candidate for theoretically formalising the process of moving symbolic and performative contents between different contexts. Simple repeated exposure does not seem to be at all sufficient to modify human behaviour. At the same time, while we highlight the absence of a solid theoretical framework, we also observe on a daily basis the acceptance of sexist, patriarchal, or immoral behaviours. In general, we notice how the introduction of symbolic content into societies and individual behaviour can act as a facilitator for the future normalisation of these mechanisms.

From the point of view of psychological mechanisms, an interesting framework to account for this theoretical issue is that discussed in Massa et al. (2022). We think it might be interesting to briefly go over the main points of this paper in order to introduce the problem of the robot's intersubjective status.

The work starts by asking what exactly the mechanism is that leads to symbolic displacement. The criteria identified are essentially two: the degree of immersivity of the experience and the degree of collusiveness of the artefact. The first criterion reformulates in a more precise way the concept of "object mediation" (Bisconti, 2021). The degree of immersivity is a measure closely related to research on augmented reality and virtual reality technologies, which shows that this type of technology can be used to create an immersive experience (Glantz et al., 1997; Riva & Waterworth, 2003; Schultheis & Rizzo, 2001). This happens because AR and VR allow the training of certain sensorimotor responses in virtualised environments, even proving useful for psychotherapy in dedicated therapeutic settings. What is enabled by VR and AR is the generation of a metacognitive process more intense than that provoked by memory or visualisation techniques. This result is in line with the idea that individual experiences are cognitively embodied and, therefore, that the same sensorimotor circuits are activated by a simulation as by the corresponding reality when the simulation precisely reproduces that reality (Shapiro, 2019). This process could then justify the transfer process, but only in cases where immersivity and similarity to reality are extremely high. Beyond this, the degree (i.e. the frequency) of exposure to the simulated experience would be of fundamental importance in determining the likelihood of detecting a visible change within the individual's interactional or behavioural dynamics. These conclusions are also in line with what has been observed by Rosenthal-von der Pütten et al. (2014) on brain activation by fMRI.

The second concept, that of collusiveness, concerns not the transfer process but its content. It identifies the relational posture that the companion robot, and the sexual robot as a striking example, implicitly and automatically establishes with the user as a full validation and confirmation of the relational content brought by the human user. But what makes the robot collusive? The following pages are concerned with developing an account, through a psychoanalytical approach, of the functioning, economics, and effects of collusive relations between humans and anthropomorphic robots. It is useful, first of all, to start from the notion of quasi-other, developed by Ihde and taken up by Coeckelbergh, which immediately directs us towards the relational posture of the robot.

Between mirrors and intersubjectivity

In his article "You, robot: On the linguistic construction of artificial others", Coeckelbergh (2011) uses the term *quasi-other* to identify the robot's relational posture towards the user. The construction of this otherness is fundamentally a linguistic process, but Coeckelbergh specifies that he does not want to "fall" into the trap of either constructivism or "naïve" realism.

> Let me distinguish two opposing views on the relation between the social and language, which I shall name representationalism and constructivism. Both views differ from extreme idealism and naïve realism, which define the relation between language and world by absorbing the one into the other: extreme idealism (in its post-modern or structuralist version) "deletes" the world outside language; naïve realism "abolishes" the subject.
>
> (Coeckelbergh, 2011, p. 62)

In the brief discussion of philosophy of language that ensues, Coeckelbergh shows how this opposition is fundamentally a false dilemma. While representationalism claims that the social precedes language, constructivism does not claim at all that language precedes the social ontologically, with the possible exception of a few streams of radical constructivism. We have already discussed these latter streams, arguing that they end up twisting themselves into a structurally circular argument. The constructivist hypothesis on the social, on the contrary, maintains that the social (as well as the real) cannot be *experienced* outside of language. Thus, the distinction is between the ontological and the gnoseological–phenomenological levels. This distinction is well known in applied philosophy, such as Judith Butler's post-structuralist feminist approach (Butler, 1989), but the constructivist position on the social is still often misunderstood. However, we hope that the previous argument on the mechanics of crises of sociotechnical systems, which comes from a markedly constructivist approach, succeeds in showing that materiality is not ultimately dissolved into language. Against this backdrop, however, we agree with Coeckelbergh's conclusion that language is a mediator that is always present in the experience of the social.

And so the SR, too, cannot be experienced outside the sphere of linguistic construction. An example is the difference between "it" and "you" when talking to the robot (Coeckelbergh, 2011 p. 62).

> Interaction based on this appearance constitutes a (quasi) social relation between us and the robot, regardless of the robot's ontological status as defined by modern science and by traditional and modern metaphysics, which view the robot as a mere thing or machine.

The result is a quasi-social relationship with a quasi-other, the robot, whose otherness is experienced through linguistic mediation. Coeckelbergh takes this concept from the already cited Ihde, who makes it explicit in his seminal essay "Technology and the Lifeworld" (Ihde, 1990), where he speaks of three ways in which technology is experienced. After *background* and *hermeneutic*, there is *quasi-otherness:*

> What the quasi otherness of alterity relations does show is that humans may relate positively or presentially to technologies [...]. Technologies emerge as focal entities that may receive the multiple attentions humans give the different forms of the other. In alterity relations there may be, but need not be, a relation through the technology to the world [...]. The world, in this case, may remain context and background, and the technology may emerge as the foreground and focal quasi-other with which I momentarily engage.
>
> (Ihde, 1990, p. 107)

Ihde's quasi-other is thus no longer instrumental, as in *embodiment* or hermeneutic relations in which technology mediates the user's relationship with the world, but instead appears "as an other to whom I relate". Is this otherness, the one Ihde speaks of, the kind of otherness we refer to in intersubjective relations? Certainly, the object can appear in its otherness when it escapes the instrumental categories of embodiment and hermeneutics.

The emergence of otherness, however, does not imply that it is possible to trace the emergence of the intersubjective relation: the object, even when it appears to the subject in its structural otherness and even if it appears in the form of a total alien (Coeckelbergh, 2016), remains inert and does not in turn recognise the human subject (Ramey, 1998).[3] We have analysed elsewhere (Bisconti & Carnevale, 2022) the relationship between alterity and intersubjectivity. There we discussed a book of capital importance for the question of the subjective-relational status of the SR: *The Changing Face of Alterity* (Gunkel, 2017). In that book, contributions by David Gunkel and Coeckelbergh analyse the problem of how technologies can change the way we see and experience otherness.

In Gunkel (2017) the problem raised by "machine otherness" is clearly stated:

> The problem with our socially situated and increasingly interactive devices is not that they substitute a machine interface for the face-to-face relationships we used to have with others. Instead, it is in the face of the machine that we are challenged to re-examine critically who or what is, can be, or should be Other.
>
> (p. 198)

The argument we made in Bisconti and Carnevale (2022) was fundamentally about the difference between the emergence of otherness and the possibility of intersubjectivity. The latter is only possible given the mutuality of recognition in a Kojevian fashion, which distinguishes relations with objects from those with subjects. In Kojeve (1980), building on Hegel's phenomenology, mutual recognition is the "desire for the desire of the other". It is this structural necessity of the other that constitutes the ground for intersubjectivity. The Kojevian theoretical background could turn out to deviate from the a-anthropocentric objectives that have led us here. In fact, the conceptual structure of recognition requires the acceptance of a clear division between object relations and intersubjective relations and therefore of a division between human and non-human. The return of the structural difference between object relations and intersubjective relations, however, should not be taken as an analytical horizon but rather as a surreptitious standpoint towards the design of social machines. What we mean here is that the constant tension towards anthropomorphising the SR reveals the assumption that sociality is only possible with anthropomorphic beings. The mutuality of recognition is thus necessary to the relationship on the condition that we follow the anthropocentric assumption that guides the design.

We know from Krämer et al. (2011) that the more anthropomorphic the robot, the more humans will adopt anthropomorphic interactional styles.

This anthropomorphic setting presupposes that the interactive machine is actually able to carry out a human-like relationality. From the perspective of Kojevian recognition theory, this means that, in the relationship with the machine, some otherness must be able to emerge. Otherness is in fact what would distinguish an intersubjective relationship from an objectual one. In this context, it matters little whether this otherness is the product of intentionality, consciousness, free will, or any other metaphysical support for the adjudication of ontological differences between humans and other entities. What the alterity experienced in the intersubjective relation does is to "dislocate", to create internal movement in the subject, who suddenly realises that on the other side of the relationship there is an "other" subject, irreducible to the gnoseological categories of objectivity. This irreducibility is the otherness that the subject is supposed to face in human interaction; the reference here to the Hegelian master–servant dialectic is clear. Even if we were to admit the validity of this approach assuming a clear division between human and non-human interactions, which we have already broadly criticised, the robot is still not in a position to manifest this otherness. As we also discussed earlier, in the case of an anthropocentric design, the occurrence of an "altering" event, of an emergence of an "otherness" of the robot, would only be possible, if at all, in the case of an error in behaviour, of an unplanned event. In this respect, we would like to suggest a parallel, which will be taken up later in the conclusions, between relational otherness and mediation as understood in ANT. Both concepts, in fact, convey a single content: that some interactions modify the systemic balance of autopoiesis, while others do not. In Coeckelbergh (2016), it is clear that the emergence of relational otherness can occur when the other is not reduced to the subject's gnoseological categories, when it is not already normed within them. From a phenomenological point of view, we could say that the emergence of otherness has as a necessary, though perhaps not sufficient, condition the occurrence of something unexpected, not gnoseologically pre-categorised.

Designers of anthropomorphic robots thus promise what they cannot actually deliver, namely the emergence of a structure of otherness within the relationship with the human being. This promise is conveyed by the hyper-anthropomorphic relationship structures themselves, including the mimicking of gestures of affection: the purr of zoomorphic robots, the simulation of sexual enjoyment by sexual robots, the reciprocation of the user's words of affection by MyRealDolls. Alternatively, it is delivered through the robot's non-verbal interactions, such as an anthropomorphic (and not strictly functional) use of the actuators. This hyper-anthropomorphism is an adaptation to the human gnoseological categories of interaction, i.e. an attempt to meet as closely as possible the human expectation of an anthropomorphic interaction. It is precisely because of the tendency to adhere as closely as possible to these expectations that the emergence of an alterity on the part of the robot in the interaction cannot in fact be provided: the robot is incapable of manifesting an anthropomorphic otherness because it does not actually manage the metacommunicative and other relational planes that could produce a significant mode of otherness and therefore of mediation in the interaction. Nor can the robot provide a "allomorphic" otherness, as it is anthroponormed, "*by-design*", according to a fashionable formula in technology ethics circles. Here we can draw the first line of conclusions, therefore, regarding the intersubjective structure of the human–robot relationship: the anthropomorphism of the robot and the anthroponormativity of its design have the double characteristic of promising relational otherness and annihilating the possibility of its emergence. The promise is given by the fact that intersubjective otherness is typically experienced with human beings in the anthropocentric world, and the robot is intended to mimic the social and relational characteristics of this otherness as much as possible. The impossibility of this emergence is given by the fact that the robot cannot, at least at the current stage of technological evolution, manage the systemic–semiotic complexity of human metacommunication in social interactions. The robot presents itself as a relational "other" but cannot bring a relational content within the relationship with the user as it does not manage the metacommunicative level. As in the case of the elderly Andy when he projects the figure of his ex-wife onto the robot (Turkle et al., 2006), the robot is not able to adapt its behaviour to the "unwritten" rules of human relationality and reformulate the interaction. Similarly, the sexual robot is not able to modulate, and maybe even interrupt, sexual intercourse in the event that the user simulates violence. These are two simplified examples to illustrate why the robot is not able to manage the contextual variety of the relationship between the action and its meaning for the subject. For this reason, we can call the robot a relational mirror. The robot responds to the subject by confirming and colluding with the subject's unconscious expectations of the interaction. As a result, the user is trapped in a completely egoic relational universe. A narrative representation of this process can be found in the film *Larry and a Girl of His Own*, in which the protagonist hallucinates his relationship with a doll, believing (perhaps only partially) that he is interacting with a real girl. The interactive structure of the robot can therefore only increase the hallucinatory nature of the mirroring relationship with the quasi-other.

What is the function of the structure of otherness that the human subject has assumed in interaction? That of a dislocated space of symbolic production of the subject's own self. In other words, the zone of otherness presupposed by the user due to the high degree of anthropomorphism in the relationship cannot be managed by the robot: the zone of otherness is made meaningful by the contents that the human puts into the relationship, which take on their meaning in the semantic–relational mirroring relationship that the robot produces with the user.

Before descending into a theory of object relations in order to further specify the meaning of "collusive relationship", let us take up a formula that summarises the nature of the human–SR relationship as set out in another work:

human → SocialRobot (~~alienated-self~~)

(Bisconti & Carnevale, 2022)

This formula summarises the fact that the human being, in turning towards the robot, experiences as otherness his or her subjective projections alienated in the interaction. The reason these are alienated relates to the robot's collusive interactivity, which reminds the subject to be sympathetic to the content of the projection when in fact the robot cannot understand or manage its implicit relational content. The bar thus identifies the elision of the relationship between the subject and the subject's own projective contents, which are dislocated and can no longer be traced back to the self.

The more anthropomorphic the robot, the higher the degree of intersubjective otherness perceived by the user. Since the robot does not carry an autonomous relational content, the otherness perceived by the user ultimately turns out to be a projective content. The relational posture of the robot, unable to manage the metacommunicative aspects of the interaction, is that of an other-mirror, which actively validates the user's projective content. This validation process is at the basis of the collusive nature of the SR. The formula, therefore, formalises the fact that the human being turned relationally toward the robot experiences in that relationship his or her own projective contents, alienated and displaced onto the machine, which actively interacts to confirm and validate them.

This process is not dissimilar, for example, to that of a child who imaginatively animates a doll and interacts with it as if it were alive. However, the doll cannot actively collude with the child's relational fantasy as it is not able to interactively validate projective fantasies. The distinctive functioning of the robot, its confirmatory and validating interactivity, can in our opinion modify structural elements of the relational organisation of individuals. The robot is a quasi-other that is placed in a grey area between object relations and intersubjective relations. In order to understand the functioning and the effects of this collusive and confirmatory relationship, we must therefore analyse the relationship and the difference that exist between these two types of relations and understand whether interaction with a robot actually lacks the elements that characterise intersubjective relationships from the point of view of relational organisation. It is within the mechanism of object relations that we want to fully investigate this process.

Internal objects, external objects

At the end of the 1960s, Donald Winnicott (1969) published a manuscript titled "The Use of an Object", in which he formalised the difference between relationship through identifications and the subsequent relation of use in the infant, a distinction and progression that marks the passage from the infantile to the adult modality of interacting with the external world. This passage eventually leads to a relational organisation capable of intersubjective relations. The first, the relationship through identifications, is characterised by the lack of distinction between what Winnicott calls the internal object, the inner representation that the subject makes of the object, and the external object, which is independent and potentially a cause of frustration. At this stage, the object is not actually perceived as external but remains a support in the environment for the subject's projective processes. Winnicott summarises the perceived external object, at this stage of individual experience, as a bundle of projections. At this stage, it is not even possible to state a logical order of precedence between the projection, the fantasy of the object, and its actual presence in reality.[4]

> One can say that first there is object-relating, then in the end there is object-use; in between, however, is the most difficult thing, perhaps, in human development; or the most irksome of all the early failures that come for mending. This thing that there is in between relating and use is the subject's placing of the object outside the area of the subject's omnipotent control, that is, the subject's perception of the object as an external phenomenon, not as a projective entity, in fact recognition of it as an entity in its own right.
>
> (p. 89)

Therefore, the fundamental characteristic of "relating through identification" is that the object is placed within the subjective zone of omnipotence. The existence of an object outside the bundle of the subject's phantasmatic projections is not recognised. In the shift to the "use of an object", the first thing that happens is the subject's attempt to destroy the object: "This change (from relating to use) means that the subject destroys the object" (p. 89). What does it mean that the object is destroyed, and why can the destruction of the object lead to a more mature relation of use? When the infant pours out aggression on the object, it is destroyed as an internal object. That is, in the infant's hallucinatory omnipotent perception, it is reduced to nothing by the subject's aggression. Yet, at the same time, the external object survives the aggressive charge and persists, resists. Winnicott continues:

> The subject can now use the object that has survived. It is important to note that it is not only that the subject destroys the object because the object is placed outside the area of omnipotent control. It is equally significant to state this the other way round and to say that it is the destruction of the object that places the object outside the area of the subject's omnipotent control. In these ways the object develops its own autonomy and life.
>
> (p. 90)

There is thus a duality in the relationship between the three elements involved: the object, omnipotence, and aggression. On the one hand, the object is attacked and destroyed because it produces frustration, and it does not conform to the subject's omnipotent fantasy. On the other hand, it is precisely because the internal object is attacked and destroyed as a bundle of projections that the external object can acquire the capacity to become autonomous as a residual survivor of the infant's aggression.[5] As in a mathematical division where "aggression" is the operator, the perception of the "surviving" external object constitutes the "residue" that cannot be further attacked. It is with respect to this residue that the relation of use is structured. The relation of use allows the object to some extent to be external in the subject's perception. The subject goes beyond relational omnipotence and proceeds towards intersubjectivity. The object thus had to undergo a phantasmatic destruction to acquire an autonomous existence as a residue. The autonomy of the object in Winnicott refers equally to human beings or inert objects, as evidenced by the fact that the first example of the transition from "relating through identifications" to the relation of use is the mother's breast.

Aggression, as a positive driving force in the evolution of the subject from an omnipotent stage to an intersubjective one, is also thematised in Benjamin's work, as for example in "Beyond Doer and Done To" (Benjamin, 2014) or the earlier "Bonds of Love" (Benjamin, 1988). Benjamin's work enquires into the conditions of the possibility of the emergence of intersubjective relationality from its objectual predecessor. She also works from a psychoanalytic background, especially leveraging the work of Melanie Klein.

In Benjamin, the intersubjective relationship emerges in a rather similar way to that described in Winnicott. The emergence of the intersubjective relationship presupposes the exit from the stage of omnipotence. In Benjamin the fundamental concept that evolves the subject out of an infantile and omnipotent relational setting is that of recognition: in an echo of Kojeve, the subject needs to find in the world the significance of actions, and this can only happen when it is recognised by another subject. The second principle that moves the human being is the principle of omnipotence over the external world, for which the adherence of the object to the phantasmatic system of projections is necessary. When the object lacks conformity, that triggers aggression. But aggression precludes recognition since it aims to remove alterity, therefore preventing the subject from being recognised by the other. The emergence of intersubjectivity is therefore only possible in a dynamic balance between the two forces, where there is no fixation in the positions of *doer* and *done-to*. Indeed, if the subject, especially in the case of the infant, remains in a condition of omnipotence, the subject will not be able to feel recognised, and the actions performed will be emptied of agency and signification. In Benjamin's words: "The parents co-opt all the child's intentions by agreement, pushing him back into an illusory oneness where he has no agency of his own". Henry (2018, p. 6) explains how there can be a positive resolution of the relationship between these two polarities:

> The recognition process occurs when the subject and the other [...] are conceived as always mutable mirror reflecting the interlocutor, so that this reflection is neither mimetic nor annihilating of the parts at play, but rather allows a continuous and permanent interchange of polarity, not fixed in an oppositional (doer/done-to) form(translation mine).

It is interesting to notice that, from a theoretical point of view, both Winnicott and Benjamin make intersubjectivity emerge from a completely selfish position of the subject. This avoids the need for major theoretical efforts to call into play "moral" justifications for the emergence of intersubjective relations and ultimately therefore also of care. We believe, however, that Winnicott succeeds in giving an even more precise account of the emergence of the intersubjective relation from within the object relations themselves. This is represented by the concept of residue. In contrast, for Benjamin the emergence of the significance of one's own action takes place only in relation to the recognising other.

The theoretical background of object relations thus allows us to identify the structure that organises the subject's relationship with the object. At the basis of the use of an object, in the mature relational organisation, there is the emergence of a residue of the external object, which resists and survives the aggressive charge. One therefore relates to the object through this residuality.

Can the robot produce residuality? We are finally in a position to answer the question of this chapter, namely what kind of relational setting the robot enacts in its interaction with the human being. The fundamental problem of HRIs is that no object residuality can be produced in the interaction as in no case does the robot resist the "aggressive discharges" enacted by the subject. We do not mean, as with the infant, an actual destruction of the object, but rather a relational reduction of the otherness of the other to projective contents. Current SRs do not perform any kind of semantic negotiation in the relationship: the horizon of meaning of the relationship is therefore completely dictated by the interacting human being. We can now specify the concept of collusiveness outlined in the previous chapter. Collusiveness describes an external object that, to some extent, becomes internal, adapting itself as much as possible to the subject's projective semantic contents, progressively bringing the user back to the stage of omnipotence. The SR, as a relational object, returns to being nothing more than a bundle of projections of the subject's semantic–relational contents.

We have identified, in the theories of Winnicott and Benjamin, the object residue as the support of the intersubjective relationship with the world. This residue-remnant accounts for the concept of "otherness" in relational experiences: that element exceeding the normative gnoseological categories into which the subject attempts to tame the real. However, it is still not clear how the residue functions and how the relationship between the subject and the residue is organised. In the next section, we discuss the function of the residue under the concept of *objet petit a*, a concept from Lacanian psychoanalysis. Finally, this chapter will end with a discussion of the "small object *r*": an attempt to understand what kind of support the robot is.

The *objet petit a*: excess and residue

Jacques Lacan's work is notoriously difficult, often non-linear and fragmented. In spite of the many difficulties with interpreting his thought, there is a certain agreement that the three main elements of the French psychoanalyst's contribution can be summed up essentially as the imaginary, the symbolic, and the Real. The contribution of this manuscript ventures considerably beyond the usual field of application of the Lacanian approach, applying its ideas to the field of human–machine interactions. Let us therefore make a caveat from the outset: the use that this work makes of Lacan is not intended to be a "scholastic" and faithful application of his teachings. On the contrary, we aim to show how some ideas from this psychoanalytical approach can systematise our understanding of human–machine relations, if considered in an anthropocentric way.

The notion we will underline is that of the *objet petit a*, often translated as "unattainable object of desire", understood in Lacan as the phantasmatic support of desire. In contrast, as already discussed above, the Real accounts for a "residue" that cannot be further symbolised. In the previous pages of this manuscript, we have already underlined how Lacanian notions of the Real differ from those of common language. It is useful to repeat this notion: the Lacanian Real identifies the structural difference between the (ontological) being and the gnoseological categories. The human symbolic structure manages to include in the realm of significance only a part of the Real, only some modes of its organisation and relationship. The others are outside the framework of language. This should not, however, lead us to adopt a kind of metaphysical dualism, as in object-oriented ontology (OOO). The Lacanian Real is always "lurking", ready to show itself to the subject disarticulating the current interpretative structures. Since the beginning of this manuscript, the topic of anguish in relation to the *ens*, not yet normalised by symbolic–linguistic categories, has periodically re-emerged. The *"Real"* is thus opposed to the *"real"* to the same extent that the *ens* is opposed to the object. The distinction we make in Lacanian terms between entity and object is reformulated, in German, as the distinction between Sache and Ding. Sache is the "thing" product of the work of human action; it is normed within the gnoseological categories of subjects and is accessible in symbolic form. Das Ding, in contrast, falls within the Real. Obviously, Das Ding is a theoretical object: in fact, only the Sache is experienceable, since only it can be symbolised. Das Ding is a representative object of the structure of the Real. It is in the realm of the pre-symbolic, of the not yet symbolised, and presses on the interpretative structures of the subject, modifying and shaping them. The Real in itself is therefore in effect not experienceable by the subject, since the subject always experiences it in a way mediated by the symbolic. It therefore represents a ὕλη, an irreducible persistence of the material with respect to the symbolic. The Lacanian subject, however, can only create experience within the symbolic, within a linguistic structure (understood here in a very broad sense, almost co-extensive with that of "signifier"). Outside of this, there is simply no experience, only pure "evenementiality". The subject experiencing the Real, in fact, does not exist: there would no longer even be a subject at that point. The subject relates exclusively to whatever of the Real has been signified, that is, has entered into the symbolic of the subject. We underline how this relationship between the human subject and the Lacanian Real precisely reflects the mechanism we discussed regarding the socio-symbolic organisation. The same problem was experienced by autopoietic units and socio-symbolic networks of actors: the main objective of both the autopoietic unit and the Lacanian subject is to tame the external world when it appears outside of language as pure Event. Therefore, the Lacanian framework we are discussing here should be seen as a specification of the above discussion on the functioning of the Real. The main difference is that in this case we are analysing the issue at the subjective, psychological level; in the previous chapters we were discussing it at the socio-symbolic level. In Lacan, the not-yet-symbolised Real is perceived by the subject in the two forms of desire and anguish. The way in which the Real comes to the subject is through the *objet petit a*, understood as an "excessive residue". It represents a "reference" pointing to an empty space, namely the not-yet-symbolised. For this reason, in relation with the Real, it represents a "residue". Conversely, in relation to the symbolic, it shows itself as an excess, a surplus, since it cannot be symbolised. The *objet petit a* is fundamental to Lacanian theory since it is the "internal engine" of the relationship between the subject and the external world, including intersubjective relations. In the following pages, we outline how Lacan gives an account of subjective experience, relating it to object relations theory. This will clarify the role of the *objet petit a* as a support of desire and as a mediator of the subject's experience. Lastly, we comment on the structure of the human–robot relationship within its current anthropocentric premises. Chapter 7 will be concerned with outlining a post-anthropocentric approach to the design of human–machine interaction.

The phantasmatic structure of experience and Das Ding

What are we led to articulate the apparatus of perception onto? Onto reality, of course. Yet, if we follow Freud's hypothesis, on what theoretically is the control of the pleasure principle exercised? Precisely on perception, and it is here that one finds the originality of his contribution. The primary process [...] tends to be exercised toward an identity of perception. It doesn't matter whether it is real or hallucinated, such an identity will always tend to be established. If it isn't lucky enough to coincide with reality, it will be hallucinated. [...]

On the other hand, what does the secondary process tend towards? [...] An identity of thought. What does that mean? It means that the interior functioning of the psychic apparatus [...] occurs as a kind of groping forward, a rectifying test, thanks to which the subject [...] will conduct the series of tests or of detours that will gradually lead him to anastomosis and to moving beyond the testing of the surrounding system of different objects present at that moment of its experience. One might say that the backcloth of experience consists in the construction of a certain system of Wunsch or of Erwartung of pleasure, defined as anticipated pleasure, and which tends for this reason to realize itself autonomously in its own sphere, theoretically without expecting anything from the outside. It moves directly toward a fulfillment highly antithetical to whatever triggers it.

(Lacan & Miller, 2013, p. 31)

That's what Freud indicates when he says that "the first and most immediate goal of the test of reality is not to find in a real perception an object which corresponds to the one which the subject represents to himself at that moment, but to find it again, to confirm that it is still present in reality".

(Lacan & Miller, 2013, p. 52)

These brief excerpts from Seminar VII clarify Lacan's position regarding the structure of human experience very efficiently and with a clarity that rarely characterises him. The Freudian dyad between the pleasure principle and the reality principle is understood by Lacan as based on a structurally hallucinatory experience of the world. The pleasure principle, which tends towards discharge, towards the satisfaction of the impulse, does not enter into a relationship with reality, but rather first of all hallucinates the satisfaction of desire. What brings into play a relation with the real, here understood simply as the external object, is the secondary process. This tries to keep the subject in a certain balance with the external world by sifting through it in search of confirmation, external and objectual, that the desire has actually been satisfied on an object that could be perceived. If this object is not found, the satisfaction and the perception have been hallucinatory. We notice two issues emerging from this excerpt: first, perception cannot enter into a non-hallucinatory relation with the real, despite the fact that it is in fact in charge precisely of relating to the *iletic*, sensory data. What manages the relation with the real is the principle of reality, which has the task of "balancing the books" of the hallucinatory structure of perception. Second, this reflection by Lacan should bring to mind what Maturana and his co-authors argued in the manuscript on frog perception (Lettvin et al., 1959): what the frog's perceptual system perceives is not the real, the external object "in its purity", which could then be received by the cognitive system. On the contrary, it is the perceptual system that conveys information to satisfy the pleasure principle – that is, it responds to a structure of desire: in the case of the frog, it only sees objects that move quickly because the frog wants to eat flies. What Lacan suggests in this passage does not therefore seem particularly different from Lettvin and Maturana's study, characterising the functioning of perception in a constructivist way.

Maturana, however, conceives of a completely solipsistic subject who endo-generates the elements of his own autopoiesis but never in fact enters into a relationship with the Real. In short, if the Real did not exist, Maturana's subject would live better, being able to maintain internal homeostasis with more precision. The Real, in Maturana, is elided by the subject, does not generate any residuality and does not enter into any effective relationship with the subject. The structure of the subject's relationship with the world as depicted by Maturana traces an unbridgeable distance between the subject and the world. What is puzzling is that the relationship between these two is ultimately configured as extrinsic. Nothing links the autopoiesis and the realisation of the living being, in the sense of the production of its components, with the Real. They are related simply by a physically necessary relationship: the subject and the Real eventually live in the same place. In line with what Hayles argues, Maturana's subject is solipsistic, and the world is essentially external to the subject, experienced as an "accident" to be managed through autopoiesis, which refers exclusively to itself. In contrast, for Lacan, there is a very close relationship between the subject and the world. This relationship passes through the path of desire, which structures the subject's relationship with the world. Desire, since it is in intimate relation with the Real, makes the subject dependent on the Real – in Lacan's words, "a barred subject". Lacan is an interesting theorist for conceptualising the relationship between the subject and the world: he relentlessly attempts to ground the relationship between the subject and the Real. We believe that he also succeeds in avoiding typical theoretical loopholes in constructivism. Against this backdrop, Lacan has the merit of grounding a radically constructivist gnoseological perspective, as we have seen in the extract above, on an equally radical materialism. A radically constructivist theory, in order to remain rigorous, needs a materialist foundation, and vice versa. The passage from a world of hallucination, as described in the above passage, to an essential relation between subject and Real passes through the three notions of Das Ding, the *objet petit a* and the crossed subject. Unfortunately, the space available in this work to retrace the Lacanian framework is limited, and the topic is instrumental rather than essential to the objective of this essay. The reader will forgive the discussion's not being exhaustive.

In talking about Das Ding, one might start with a reflection: why do human beings experience incessant craving? In a homeostatic system, once the urgency of need has been satisfied, the system should return to a state of stillness and inactivity until another stimulus puts the system back into a state of disequilibrium. This is not observed in human beings, who on the contrary continue to turn their attention to the world for objects that go beyond the sphere of mere drive discharge. The idea that (i) the pleasure principle seeks to discharge, possibly in motor form, the system's arousal, externalising it, while (ii) the reality principle conforms to the social norm does not seem sufficient. In fact, this framework does not explain the incessant desire that drives humans, which rarely fixes itself on anything actually objective or on a physical drive. Even if this were the case, it would still not explain why, as soon as they have satisfied one desire, humans are already looking for another object to pursue. If the object of desire is mundane, once it is found, the system should enter a state of stillness. Yet, on the contrary, it remains in a state of constant "desiring activation". Lacan solves this problem by finding in Freud the notion of Das Ding, the Thing.[6] Das Ding is a conceptual tool to claim that human desire is structurally founded on the absence of the desired object. This, if it were ever found, would constitute the experience of *jouissance*.[7] *Jouissance*, if experienced, cancels the very experience of desiring, ultimately fulfilling homeostasis and leading the system to a state of "entropylessness", hence ultimately to death. In the seventh seminar, Lacan makes explicit the role of Das Ding in relation to the system of perception and in relation to the problem of homeostasis:

> it is, of course, clear that what is supposed to be found cannot be found again. It is in its nature that the object as such is lost. It will never be found again. Something is there while one waits for something better, or worse, but which one wants. The world of our experience, the Freudian world, assumes that it is this object, das Ding, as the absolute Other of the subject, that one is supposed to find again. It is to be found at the most as something missed. One doesn't find it, but only its pleasurable associations. It is in this state of wishing for it and waiting for it that, in the name of the pleasure principle, the optimum tension will be sought; below that there is neither perception nor effort. In the end, in the absence of something which hallucinates it in the form of a system of references, a world of perception cannot be organized in a valid way, cannot be constituted in a human way. The world of perception is represented by Freud as dependent on that fundamental hallucination without which there would be no attention available.
>
> (Lacan & Miller, 2013, pp. 53–54)

The fundamental object of desire, which if found would cause the annihilation of the subject in *jouissance*, is what enables the world of perception to follow a certain path, searching for Das Ding. This search is structurally unsuccessful and – for this very reason – keeps the system in a certain state of constant excitement. The subject fails because it seeks Das Ding within the world of the signifier, namely of the symbolic, while Das Ding actually represents its void, which orders itself according to a law not pertaining to meaningful experiences. In other words, access to the object of desire is prevented by a gnoseological barrier. Against this backdrop, within Lacanian theory we state that the subject structurally loses access to the object of the desire at the moment that s/he enters the symbolic, when becomes a linguistic being. Nevertheless, Das Ding remains, for the subject, always dependent on the Other. In the following Figure 6.1, the division exemplifies the relationship between the subject, the Other, and the *objet petit a*.

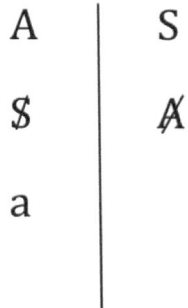

Figure 6.1 Graphic representation of "a" as a residue.

Source: Lacan, 2014.

In Lacan, the Other is both (i) the locus of desire and (ii) the source of the signifier, which produces the bar on the subject. As (i), it makes explicit the function of Das Ding, which is structurally outside the subject (even of its gnoseological capacity). The subject will search for Das Ding in the external object, the Other. Moreover, the reason for the Other's role as (ii) is traceable in the theory of the mirror (Lacan, 1974). This shows the structural dependence that the subject has on the Other with respect to the universe of the signifier. It is a common experience to note the amazement that children less than a year-old experience when looking at their own reflection in the mirror. According to Lacan, this amazement concerns the fact that the mirror provides the subject, for the first time, with a unitary image of the self. Before this reflected experience, the infant's self-perception consists of a series of proprioceptive sensations of the body. However, these are not experienced as a unity but rather as a fragmentary set of partially controllable body parts. In fact, even from a physical–motor point of view, the infant is not completely in control of his or her body. At the moment when the infant is reflected in the mirror, this "body in fragments" achieves a representable unity. In Lacanian terms, what is before the mirror is the *je*, the subject's unreflective self-perception. The image in the mirror makes it possible to recognise a *moi*, a reflexive unity. But why is the origin of the signifier in the Other? The child needs to receive from the Other the confirmation that the *moi* in the mirror is indeed a signifying unity (Lacan, 2014). The child first lives the condition of the "body in fragments", in which the unitariness of her/his own limbs is not yet acquired (at this stage, the child is therefore a *je*). Then s/he observes in the mirror a unitary and coherent figure, an image that represents her/him as not fragmentary. What the child then does, after a moment of astonishment, is turn to the Other (usually a parent) to obtain confirmation of that unitary image, which is of the order of the signifier. The subject as reflexive and unitary *moi* is an imaginary reconstruction of the fragmented body, possible only through the "guarantee" of the Other. The entry into the world of the signifier "places the bar" on the subject, testifying that something has been lost: the subject is now in structural dependence on an Other who is the guarantor of the signifier. Returning to the equation found in Seminar X (the image above), on the right we find the subject before the entry into the signifier and therefore without the bar; on the left we see the great Other, guarantor of the signifier. These produce the barred subject, who no longer has access to a self that is not reflexive, and therefore symbolic, in the form of the *moi*. This reflection entails dependence on a signifier that does not belong to the subject, but of which the Other is guarantor. The Other in turn bars itself: the subject will never find Das Ding in the Other. From the Other, finally, a residue is generated, *a* or *l'objet petit a*, which is that produced by the equation.

In the end, the equation explains how subjects enter the realm of signifier. The *a* prevents the subject from going "back" to the stage of a relationship with a Real not mediated by the Symbolic. Obviously, as we have already explained, this "going back" is not really a possible action and only identifies a logical precedence of the Real with respect to the Symbolic, which has no reference in the experience of the subjects. Yet, as in Benjamin, only through the Other can we search for the lost object, ultimately to discover that the Other itself is not at all a good guarantor of our desire, since it is not Das Ding. From this series of relations between the subject, the Other, the signifier, and the Real we can understand the double, Janus-faced nature of the *objet petit a*. This is, in fact, at once a trace, a residue, and an excess, a surplus.

The *objet petit a* occupies an important part of Lacan's thought in his attempt to conceptualise the structure of human experience. The function of *a* can be summarised as a support of desire. The *objet petit a* represents the coordinates of pleasure that guide the system of attention, the pleasure principle of the subject. S/he is in search of *jouissance*, that is, a totalising and annihilating satisfaction of her/his own desire. Against this backdrop, *a* represents that trace of the Real which, not being symbolisable, stresses the subject's experience. This pressure of the Real has two effects: first, it reminds the subject that *jouissance* is unreachable – here is the function of the residue. Second, it testifies to the structural insufficiency of the symbolic in signifying the subject's experiences. In fact, this desire is structurally impossible to satisfy. Here *a* has the function of an excess. It is a residual remnant of the Real, evidence of something non-symbolisable in human experience, something that exceeds the dimension of the signifier. The residuality of *a* itself produces desire, which, unlike *jouissance*, is always a pleasure *in the signifier*. How, however, does *a* cause the desire to move in the realm of the signifier from the inaccessible *jouissance* of the Ding? Desire cannot be experienced except in the world of the signifier: this is the only form in which humans can access desire. What is important to understand is that *a* in Lacanian theory is not an object in itself, experienced as Real, but a characteristic of meaningful external objects. "*a*" represents a trace of Das Ding left on experienceable objects: it testifies to a void at the centre of the object.[8] The *objet petit a* therefore allows the slipping of desire into the world of the signifier, since it is a residue of the Real – Das Ding. Compared to the object of experience, *a* is also a surplus: the subject searches in the object for a surplus of pleasure, the *jouissance*, which s/he does not actually find. It is not actually findable, since that surplus pertains to the Ding. Finally, the object of desire is not the object itself but its surplus, which refers to the Real, the "little nothing" that is sought within it. The conclusion of the discussion is that the *objet petit a* is the phantasmatic support of desire, which allows desire to move from object to object, of which it represents a residue and a surplus.

Before zooming in on how *a* works in the case of HRI, we want to dwell on some considerations about Lacanian theory. We believe it is clear at this point how, from a theoretical point of view, the above approach by the French psychoanalyst substantiates, and largely surpasses, Winnicott's conclusions regarding the use of an object. Moreover, through Lacan's thought we are also better able to analyse the concept we underlined in Benjamin, namely the emergence of the significance of one's own action only in relation to the other. Lacan takes both thoughts much further with the fundamental intuition of the notion of the Real. From this perspective, the concept of an intersubjective relation is completely reshaped. The Other of the subject is structurally unattainable. It fluctuates between being the guarantor of the symbolic, the Big Other, and the residual excess *a.*, the small other. As a result, the subject's relationship to the world is constantly mediated by a hallucinatory experiential structure. The relationally "healthy" subject has the capacity or luck to find a match between the hallucination of the desire and the external world. If on the one hand Lacan precludes the subject from a fully intersubjective relationship, on the other hand he demolishes any claim to placing the relationships between human beings at the centre of the universe: the relationship, whether objectual or intersubjective, is always mediated by a phantom, a support of desire. The Lacanian subject, however, unlike Maturana's subject, is not completely enclosed in a solipsistic structure of relation to the world: the *objet a*, and desire in general, organises the subject's experience in *structural dependence on the world of experience*. Despite the obvious limitation of the Lacanian approach as a potential theoretical background for a non-anthropocentric theory of the subject–world relationship, i.e. one that is not based on the analysis of the experience of the human subject alone, we believe that the advantages it offers remain valuable. The deconstruction of the ethical priority of the intersubjective relationship puts us in the condition of thinking about relations between humans and the world that are structured asymmetrically. The structural dependence of the subject on the world, the Real, on the one hand attenuates the strong constructivism of a hallucinatory theory of perceptions and on the other demolishes the claim to autonomy of the subject with respect to the universe of being. Finally, the notion of the Real opens a rift in the process of domestication of the entity in question, which we can now recognise as a structurally unsuccessful enterprise. We can also finally recognise the scope that guides socio-symbolic systems in their organisation: the Real is a cause of anguish when it is not symbolised and gnoseologically tamed. The time has come to draw conclusions about the relational posture of the anthropomorphic SR, at which the reader's intuition may have already arrived.

What kind of object is the robot? What kind of desire organisation does it support? The functioning of the *objet petit a* leads us to some considerations about the effect of such an object on the structure of human experience. The positive function of *a* is to structure the relationship of the human with the world in a constant dimension of *estimity, a* neologism that identifies something as close to us while at the same time maintaining a form of extraneity. This relationship is creative insofar as it guarantees the subject a constant relationship with the alterity. Certainly, this otherness can only be experienced in the form of the signifier and is therefore gnoseologically normative. On the other hand, the reduction of the Ding to a Sache does not exhaust the intrinsic excess of the Real over the Symbolic. This ensures the persistence of an excessive residuality. This is the form of a possible intersubjectivity, the way out of a solipsistic condition into which the hallucinatory human monad would otherwise fall. In its encounter with the residue, the human system needs to continually adapt its enunciative universe, namely the set of signifiers that for that subject constitute the horizon of the world's meaning. Against this backdrop, the claims made about the structures of socio-symbolic systems can be transferred to the structure of individual experience: the reorganisation of the subject's enunciative structures flows in the chain of the signifier and is reorganised in such a way as to give meaning to this residue, constantly moving target. The anthropomorphic, anthropocentric, and anthroponormative SR diverts this process, structuring it instead in a solipsistic key. The *objet petit a* is the result of a division between S, the subject, and A, the other. If the "other" is actually a mirror of the subject, the access to otherness is structurally precluded. In fact, the subject will look for a residuality where it cannot find it, since the anthropomorphic-other-robot is not a signifier that produces a trace. The robot is constituted only by the projections–identifications that the subject produces in the relationship. Where *a* could be produced, opening the way to a relationship with the Real and with a structural otherness, with the robot a process of subject alienation is produced. The *petit objet r(obot)*, the no longer residual product of the relationship with the robot, is the set of projections produced by the subject and alienated from it in a surreptitious and unconscious way.

> the first and most immediate goal of the test of reality is not to find in a real perception an object which corresponds to the one which the subject represents to himself at that moment, but to find it again, to confirm that it is still present in reality.
>
> (Lacan & Miller, 2013, p. 52)

There is no longer any doubt that the subject will rediscover and ascertain the existence, outside the self, of the hallucinatory perception that guided the primary process, since there is no longer anything other than his or her own relational hallucinations to substantiate the significant elements of the relationship with the SR.

With the support of different theoretical instrumentation, we have already supported this hypothesis in other words in another work (Bisconti & Carnevale, 2022):

human → SocialRobot (alienated-self)

This formula illustrates that what the subject finds in the interaction with the anthropomorphic machine is the projective content of her or his fantasies, alienated and experienced as otherness.

When the human relates (→) with SRs, the relational content found in that relation is not an "*a*", the emergence of an otherness, as when relating with the external world. The relational contents experienced are instead the relational projections of the human, surreptitiously alienated in the machine. Otherness is therefore ultimately precluded.

In conclusion, the current anthropomorphic and collusive structure of the design of SR interaction constitutes a solipsistic and structurally hallucinatory relationality. The SR, characterised in this way, precludes access to a relationship with otherness. It rather presents self-referential relationality in an egosyntonic form for the subject, since it cannot be experienced as alienation.

This chapter started with the purpose of analysing what effects SR, in their current anthropomorphic design, might have on humans. This issue was divided into two questions. The first question was related to what mechanism enabled this displacement, and under what conditions. The second question concerned the relational content that would be transferred from human–machine interactions to those between humans. Regarding the first question, we outlined within the theory of *embodied experiences* how the process of transfer requires an immersive condition, which can activate the same sensorimotor responses in the two contexts. Regarding the second question, we had preliminarily identified the content in the concept of the robot's collusiveness. This, in an immersive condition, would imply an expectation of relational collusiveness in interactions between humans as well. In order to fully understand the implications of this collusiveness, we analysed its psychological function within *object relation theory* and Jessica Benjamin's approach. The collusive robot provokes a regression of the subject's ability to differentiate between external and internal objects and elides the relationship with object residuality. We have explored this concept of residuality in greater depth through the Lacanian *objet petit a*. This eventually led us to define the anthropo-normed robot as a phantasmatic support. We concluded that current anthropomorphic HRI produces an alienation of the projective contents of the user, who experiences them as relational otherness. This process finally locks the user into a hallucinatory and solipsistic relationality.

This relational implication of SRs, of course, will not manifest itself in all users and with the same evidence. As we discussed at the beginning of this section, some users of SRs, and especially of companion robots, may nevertheless have a limited social life. In such cases, it is presumable that the psychological effects we have discussed may emerge more strongly.

The discussion we have just concluded concerned the relational implications of SRs conceived within an anthropomorphic and anthropo-centric design, in which the goal of the robot is to mimic human inter-actional characteristics as precisely as possible.

Next section will present a research agenda on "hybrid sociality" of social systems. Before digging into these issues, we draw some preliminary conclusions on the work done so far.

Preliminary conclusions

This work discussed the possible effects of SRs for our societies. In order to accomplish this task, we had to take seriously the notion of socio-technical systems and ground this notion on a solid theoretical framework. To ana-lyse technology impact always means to give an account of what social systems are, and how do they work. Therefore, for some pages we had to higher the scope of our enquiry to understand how social systems embed technology in their functioning, in their autopoiesis. What we discovered is that we have always been hybrid, that non-human actors have always already mediated sociality between humans. Actor–network theory (ANT) offers an interesting framework to consider the role of non-human actors. We built on top of ANT a theoretical framework to understand at what condition an actor mediates the social, and at what condition an actor can bring systemic modifications to the socio-symbolic system. SRs, and in general synthetic social mediators, bring a qualitative difference with respect to other non-human actors: they enter the social sphere as "others" we can relate with. With Winnicott, Benjamin, and Lacan we enquired the intersubjective structure of relations between humans and SRs. Our conclusions suggest that anthropomorphism will lead to alienating effects on users. In these preliminary conclusions we want to discuss some theoretical and methodological issues that will bridge the previous pages with the research agenda we put forth.

To use Lacan for analysing the intersubjective implications of SRs might sound strange since the French psychoanalyst is well known for putting in danger the very concept of intersubjectivity. Apart from the *cliché* statement "there is no sexual relationship possible", the framework of Lacan's thought hardly admits to access to alterity of the other. In fact, the subject is always in some ways caged inside a hallucinatory reality, where perceptions will only (optimistically) confirm the projections of phantasy. The very relationship with the "other" is hindered and intersubjectiveness is reduced to subjects interacting with the "residual image" of the other. This is not so different from the conclusions of systemic post-rationalist psychology. The environment is set in a similar way with respect to Lacanian Real, since it is actually inaccessible to the subject, who will experience its effects as "perturbations" (Guidano et al., 2019) in the system's autopoiesis. In Lacan, similarly, the Real can be experienced only in a proxy way, mediated by the *object petit a*. It drives the subject in the world as a residue resulting from the equation between the subject's projections and the Real. This is quite a pessimistic standpoint. Intersubjectivity is implicitly hindered, at least in its full meaning, and subjects will be always locked inside themselves. What Lacan's and post-rationalist's systemic theory can show us, in their pessimistic standpoint towards intersubjectivity, is that we can still obtain better or worse "being in the world" as relational subjects. Under psychoanalytic terms, we can always be more or less in contact with the object residuality and therefore avoid excessive (and therefore patologic) projection in our relationship towards the world. We can, and we do, switch from "relating with identifications" to using objects. To preserve object's residuality is in fact exactly what intersubjectivity is, inside these frameworks. While we cannot purely access others' otherness, we can be aware of the difference between our projections and the external world. In a systemic fashion, subjects can employ different and effective strategies in order to "cope with reality", namely avoid an excessive distance between the subject's socio-symbolic organisation and the actual environment. Socio-symbolic organisation of subjects is effective when it can reliably interpret and predict environment behaviours and changes. This concerns intersubjectivity too: others (humans or not), for a system, are environment and their behaviours are perturbations that the system wants to forecast, in order to remain stable. Post-rationalist cognitive theory employs the notion of "tacit" experiences of the subjects: those are the experiences that the subjects' cognitive domain, namely its symbolic organisation, cannot access because they are not included in those experiences meaningful for the subject. To enlarge the spectrum of "meaningful" experiences allows the subject to acquire a more stable organisation, functional with the outer environment. These considerations point out why Lacan, Winnicott, and the systemic approach are all in line in troubling the anthropocentric perspective to socio-technical sociality, and the anthropomorphic design of SRs. Anthropomorphic SRs hinder the possibility for subjects to "cope" with the residuality of the Real and increase the self-referentiality of autopoietic units. The subject is caged more and more in a system of projections. What previously was a "environment", namely something structurally external to the subject, now is an alienated self. With anthropomorphic SRs, and synthetic social mediators, the environment is a part of the self. In some ways, this is nothing new: gods and idols are created to make the environment more similar to the subject's wills. If coping with the environment is too hard for the autopoietic unit, then we make the environment a piece of the self. But we should recall what happens in the Lisbon earthquake in 1755: after the earthquake, many citizens of Lisbon died; survivors took refuge in the cathedral, near the sea, to pray God for their lives. A tsunami, following the earthquake, destroyed the entire cathedral, killing the worshippers inside. As Zizek would say "there is no redemption in suffering". The same goes with SRs. To make the environment a part of the self brings the system to be highly ineffective in dealing with consequences of the outer world. In the case of SRs, it could make users ineffective in dealing with alterity relations, namely cope with environment residuality. On the other hand, ANT and system theory trouble the anthropocentric perspective toward sociality. What is social is not an agent, but a system, composed of interactions that become meaningful (and therefore social) only inside that precise system configuration. The socio-symbolic system includes and excludes interactions from being meaningful for the systemic reasons we discussed in Chapter 4. The anthropocentric perspective on the social is finally untenable. This must change the entire perspective on how we discuss, analyse, and evaluate social robotics and synthetic social mediators. Lacan, Winnicott, and Benjamin cannot give us any answer on how our analysis must change. Lacan is far from being a post-anthropocentric thinker, but in our opinion he points at the empty core of anthropocentrism. In fact, he shows how the

Real is structurally unreachable by subjects and, in the same time, how we are structurally tied to the Real. Coming back to Coeckelbergh's question "How can we know and how should we evaluate the human-robot relationship?", we understood that there must be a theoretical change to correctly interpret the role of SRs in our societies.

From the previous pages, some reader might think that we have quite a pessimistic view of technology impact on societies. We want to dispel this doubt. Many set up the arguments about technology impact in an Heideggerian fashion. I will not recall the worldwide known Heideggerian argument on *technè* and *gestell*. We do not share this approach. We claim that the radical technological pessimism fails through and through in its political scopes, for many reasons. First, it always fails in identifying the differences between the given technology that its criticised and any other technology previously designed. The worst is always "yet to come" with the next technology, which will destroy our world. While Russels' inductivist turkey eventually has a bad end, the evergreen fatalism of this kind of approach is not useful to tackle the concrete danger of emerging technologies. In fact, even if a "technology takeover" is theoretically possible, and then Russel's turkey applies, there must be a clear explanation of the reasons why *this* brand-new technology will be different from all the past technologies, in destroying our world. Between forecasting and divine events there is the huge difference of explaining the peculiarities of the *given* situation (or technology) with respect to any other else. Moreover, this kind of claims defuse their own critic potential when they take a sort of technological fatalistic standpoint: technology will ultimately destroy humanity and there is nothing we can do except refuse it all. In front of this statement, obviously no policies can ever be designed to regulate technology. These "theological luddism" hides the very fact that technology is not something detached from individuals wills and actions and is not like a Leviathan that walks on its own foot and will shape societies on its own mechanic will. Teleological theologies of technology, excuse the word play, can do anything except bring a responsible standpoint in the discussion about regulations, policies and design. On the other hand, our claim is that technologies are only other signifiers that can take on socio-symbolic meanings; and therefore technologies are mediators that can shape our societies. The notion of socio-technical systems teaches us that technology and society are a feedback-loop system: they co-interact and co-shape each other. With the notion of socio-symbolic system we have a sketch of the functioning of this co-interaction. In order to ensure that technology will be beneficial for societies, we can analyse and modify the way in which technological signifiers are signified in social systems. We suggested, throughout this work, that an anthropocentric perspective on technology is falling short in this direction: the socio-symbolic system that previously organised the interpretation of the environment cannot cope with synthetic social mediators without producing many shortcomings, as for SRs. This means that we must rethink our approach to socio-technical systems, and ultimately to design. The scope of the next section is to put forth a research agenda that follows this direction. We will discuss where can we start to adopt a systemic, ecological approach to socio-technical systems and specifically to SRs. In drafting our agenda, we start from a consideration: change in the theoretical perspective towards social robotics, and synthetic artificial agents in general, is *per se* a radical change of the object itself. If social robotics is a particular form of *creatio ex nihilo*, the creation of a social agent from scratch, the way we theorise this actor is more than ever influenced by the socio-symbolic organisation of the social system. Now we can reshape, rethink in new terms, the "anguish" that we discussed at the beginning of the book. We said that the creation of the robot, *ex nihilo* and in an anthropomorphic form, had the objective to separate as much as possible subjects and objects, to soothe the anguish of falling back in the undifferentiated entia. At the end of our journey, this means that the socio-symbolic system is trying to populate the environment with its own productions. The environment, when is an *ens*, is highly unpredictable, and generates gnoseological anxiety: the anguish of not being able to predict and interpret the environment and its effects on the system. The technological object, on the other hand, is tamed because it is a production of the system itself. Systems create subclasses of themselves to increase the internal complexity, to finally become more complex than the environment itself (Luhmann, 1995). The complexity of technology, on the other hand, is becoming overwhelming for social systems. The effects of social robotics are not as predictable as we wanted: contradicting non-verbal cues causing "uncanny valley" effect (Kätsyri et al., 2015), gaze causing anxiety (Nomura & Kanda, 2015). To avoid the anguish of the outer environment, of the Lacanian Real, we are bringing this anguish inside the system itself.

The research agenda in Chapter 7 aim to offer a roadmap to relocate the theoretical understanding of HRI, in order to radically change the design. This change should not modify HRI design understood as details in the shape or in the interaction style of the robot. Through a conceptual turn we should rethink what interactive agents are in society for: their scope and their functions might become radically different when the anthropocentric perspective is taken apart. The notion of "hybrid social systems" will be the leitmotif of our research agenda. This concept wants to recall some of the theoretical results of this work. First, anthropomorphic social interactions are a subset of social interactions. Second, that the social sphere has always been inhabited by non-human actors, mediating the social. Moreover, the fact that social robotics and synthetic social mediators will bring this consideration to a qualitatively different level, since they access the social as quasi-others. Lastly we recall that the social is a system, where all the elements (signifiers) acquire a meaning in a socio-symbolic organisation, and therefore we cannot consider interactions as untied from the social system, and we cannot consider technology as separate from its narratives.

Notes

1 This term will be reformulated in the following paragraphs through a discussion of the notion.
2 Capgras syndrome is a psychiatric illness in which the subject firmly believes that a friend, a spouse, and other family members have been replaced by impostors or impersonators – that the "copies" have appropriated the identities of people close to them and replaced them.
3 "A chair is there for my sitting, but I am nothing from the perspective of a chair" (Ramey, 1998, p. 7).
4 We will see later that the theme of the relationship between the phantasmatic object and the real object, and the problem of the precedence for the subject of one of the two, is a fundamental topic in Lacanian theory.
5 Winnicott specifies:

> At the point of development that is under consideration here, the subject goes on to create the object, in the sense of finding the "outside" itself, and it must be added that this experience depends on the object's capacity to survive (it is important that this means: "does not retaliate")

Destruction of the object, therefore, is not to be understood in the strict sense as the object no longer existing as a worldly form. More than anything else, Winnicott is emphasising that when the infant attacks the mother-object, which causes frustration by not immediately providing access to the breast, an aggressive reaction on the part of the mother does not allow the transition to the use of the object, in the same way as never frustrating the infant's desire also does not allow that transition. In fact, it is obvious that if the infant, at the moment it starts to attack, sees a condition without frustration restored (immediate access to the breast), the distinction between external and internal object cannot take place. The object has in fact come back into conformity after the attack, and therefore the aggression on the internal object has worked. The opposite, however, is less obvious: even if the mother reacts aggressively, the child's destructive action has in fact worked, because the mother-object has changed its nature. The point is therefore that the object can only be produced as a remainder if there is a clear difference between the expected effect of the child's aggressive act and the observed effect.
6 Many doubts can be raised as to how faithful Lacan actually is to the Freudian text. However, this manuscript does not aim at a philological examination of Lacan's hermeneutic accuracy.

7 On the difference between desire and *jouissance*, compare (Tarizzo, 2014) and (Lacan & Miller, 2013).
8 The *objet petit a*, in its function of testimony to a void, is quite similar to what Badiou explained about the Truth of the Being. In this case this process occurs at the level of the subject's experience.

References

Benjamin, J. (1988). *Bonds of love*. Pantheon.

Benjamin, J. (2014). *Beyond doer and done to*. Routledge.

Bisconti Lucidi, P., & Nardi, D. (2018). Companion robots: The hallucinatory danger of human-robot interactions. *AIES 2018 – Proceedings of the 2018 AAAI/ACM Conference on AI, Ethics, and Society*, 17–22. https://doi.org/10.1145/3278721.3278741

Bisconti, P. (2021). Will sexual robots modify human relationships? A psychological approach to reframe the symbolic argument. *Advanced Robotics, 35*(9), 561–571. https://doi.org/10.1080/01691864.2021.1886167

Bisconti, P., & Carnevale, A. (2022). Alienation and recognition: The Δ phenomenology of the human-social robot interaction. *Techne: Research in Philosophy and Technology, 26*(1), 147–171.

Butler, J. (1989). *Gender trouble: Feminism and the subversion of identity*. Routledge.

Coeckelbergh, M. (2010). Artificial companions: Empathy and vulnerability mirroring in human-robot relations. *Studies in Ethics, Law, and Technology, 4*(3). https://doi.org/10.2202/1941-6008.1126

Coeckelbergh, M. (2011). You, robot: On the linguistic construction of artificial others. *AI and Society, 26*(1), 61–69. https://doi.org/10.1007/s00146-010-0289-z

Coeckelbergh, M. (2016). *Alterity ex machina: The encounter with technology as an epistemological-ethical drama* (G. David, F. J, & M. D, Eds.). Rowman & Littlefield.

Coeckelbergh, M. (2018). How to describe and evaluate "deception" phenomena: Recasting the metaphysics, ethics, and politics of ICTs in terms of magic and performance and taking a relational and narrative turn. *Ethics and Information Technology, 20*(2), 71–85. https://doi.org/10.1007/s10676-017-9441-5

Danaher, J. (2017). The symbolic-consequences argument in the sex robot debate. In J. Danaher & N. McArthur (Eds.), *Robot sex: Social and ethical implications* (pp. 170–196). MIT Press.

Drummond, A., Sauer, J. D., & Garea, S. S. (2018). The infamous relationship between violent video game use and aggression: Uncharted moderators and small effects make it a far cry from certain. In *Video Game Influences on Aggression, Cognition, and Attention* (pp. 23–40). Springer International Publishing. https://doi.org/10.1007/978-3-319-95495-0_3

Glantz, K., Durlach, N. I., Barnett, R. C., & Aviles, W. A. (1997). Virtual reality (VR) and psychotherapy: Opportunities and challenges. *Presence: Teleoperators & Virtual Environments, 6*(1), 87–105.

Guidano, V., Cutolo, G., & De Pascale, A. (2019). *La struttura narrativa dell'esperienza umana*. Franco Angeli.

Gunkel, D. (2017). *The Changing face of alterity*. Rowman & Littlefield.

Gutiu, S. M. (2016). The roboticization of consent. In *Robot law* (pp. 186–212). Edward Elgar Publishing. https://doi.org/10.4337/9781783476732.00016

Henry, B. (2018). Voluntary submission as a dark side of adaptive preference. The contribution of relational psychoanalysis to Political Philosophy. *Soft Power, 9*, 99–115.

Ihde, D. (1990). *Technology and the lifeworld: From garden to earth*. Indiana University Press.

Jentsch, E. (1997). On the psychology of the uncanny (1906). *Angelaki: Journal of the Theoretical Humanities, 2*(1), 7–16. https://doi.org/10.1080/0969725970 8571910

Kätsyri, J., Förger, K., Mäkäräinen, M., & Takala, T. (2015). A review of empirical evidence on different uncanny valley hypotheses: Support for perceptual mismatch as one road to the valley of eeriness. *Frontiers in Psychology, 6*(March), 1–16. https://doi.org/10.3389/fpsyg.2015.00390

Kojeve, A. (1980). *Introduction to the reading of Hegel*. Cornell University Press.

Krämer, N. C., Eimler, S., von der Pütten, A., & Payr, S. (2011). Theory of companions: What can theoretical models contribute to applications and understanding of human-robot interaction? *Applied Artificial Intelligence, 25*(6), 474–502. https://doi.org/10.1080/08839514.2011.587153

Lacan, J. (1974). Lo stadio dello specchio come formatore della funzione dell'Io. In G. Contri (Ed.), *Scritti Vol I*. Einaudi.

Lacan, J. (2014). *Seminar X – Anguish* (J.-A. Miller, Ed.). Wiley and Blackwell.

Lacan, J., & Miller, J.-A. (2013). *The ethics of psychoanalysis 1959–1960: The seminar of Jacques Lacan*. Routledge.

Lettvin, J. Y., Maturana, H. R., McCulloch, W. S., & Pitts, W. H. (1959). What the frog's eye tells the frog's brain. *Proceedings of the IRE, 47*(11), 1940–1951.

Luhmann, N. (1995). *Social systems*. Stanford University Press.

Markey, P. M., Markey, C. N., & French, J. E. (2015). Violent video games and real-world violence: Rhetoric versus data. *Psychology of Popular Media Culture, 4*(4), 277–295. https://doi.org/10.1037/ppm0000030

Massa, N., Bisconti, P., & Nardi, D. (2022). The Psychological implications of companion robots: A theoretical framework and an experimental setup. *International Journal of Social Robotics*. https://doi.org/10.1007/s12369-021-00846-x

Mori, M. (1970). Bukimi no tani [the uncanny valley]. *Energy, 7*, 33–35.

Nomura, T., & Kanda, T. (2015). Influences of evaluation and gaze from a robot and humans' fear of negative evaluation on their preferences of the robot. *International Journal of Social Robotics, 7*(2), 155–164. https://doi.org/10.1007/s12369-014-0270-y

Peeters, A., & Haselager, P. (2019). Designing virtuous sex robots. *International Journal of Social Robotics*. https://doi.org/10.1007/s12369-019-00592-1

Ramey, C. H. (1998). "For the sake of others": The "personal" ethics of human-android interaction an unexpected ethical dilemma for android science mind-body and the natural sciences. In Proceedings of the CogSci 2005 Workshop: Toward Social Mechanisms of Android Science, pp. 137–148.

Riva, G., & Waterworth, J. A. (2003). Presence and the self: A cognitive neuroscience approach. *Presence Connect, 3*(3), 1–10.

Rosenthal-von der Pütten, A. M., Schulte, F. P., Eimler, S. C., Sobieraj, S., Hoffmann, L., Maderwald, S., Brand, M., & Krämer, N. C. (2014). Investigations on empathy towards humans and robots using fMRI. *Computers in Human Behavior, 33*, 201–212. https://doi.org/10.1016/j.chb.2014.01.004

Sætra, H. S. (2020). The parasitic nature of social AI: Sharing minds with the mindless. *Integrative Psychological and Behavioral Science, 1*. https://doi.org/10.1007/s12124-020-09523-6

Schultheis, M. T., & Rizzo, A. A. (2001). The application of virtual reality technology in rehabilitation. *Rehabilitation Psychology, 46*(3), 296.

Shapiro, L. (2019). *Embodied cognition*. Routledge.

Sharkey, A., & Sharkey, N. (2012). Granny and the robots: Ethical issues in robot care for the elderly. *Ethics and Information Technology, 14*(1), 27–40. https://doi.org/10.1007/s10676-010-9234-6

Sharkey, A., & Sharkey, N. (2021). We need to talk about deception in social robotics! *Ethics and Information Technology, 23*(3), 309–316. https://doi.org/10.1007/s10676-020-09573-9

Sharkey, N., & Sharkey, A. (2010). The crying shame of robot nannies: An ethical appraisal. *Interaction Studies Interaction Studies Social Behaviour and Communication in Biological and Artificial Systems, 11*(2), 161–190. https://doi.org/10.1075/is.11.2.01sha

Sparrow, R., & Sparrow, L. (2006). In the hands of machines? The future of aged care. *Minds and Machines, 16*(2), 141–161. https://doi.org/10.1007/s11023-006-9030-6

Tarizzo, D. (2014). *Introduzione a Lacan*. Laterza.

Turkle, S. (2017). *Alone together: Why we expect more from technology and less from each other*. Hachette UK.

Turkle, S., Taggart, W., Kidd, C. D., & Dasté, O. (2006). Relational artifacts with children and elders: The complexities of cybercompanionship. *Connection Science, 18*(4), 347–361. https://doi.org/10.1080/09540090600868912

Vallor, S. (2016). *Technology and the virtues: A philosophical guide to a future worth wanting*. Oxford University Press.

Winnicott, D. W. (1969). The use of an object. *The International Journal of Psychoanalysis, 50*, 711–716.

7 Hybrid systems

A research agenda

In the previous section, we enquired into the implications of the current anthropomorphic design of social robots. The anthropomorphic physical design of robots already highlights the direction that designers have taken. On a subtler level than this, we discussed the implications of interaction design that aims to imitate human relationality. Last but not least, the anthropocentric approach to social robotics also shapes the goal of this technology: to perfectly mimic and resemble humans in the context of companionship. Efforts in these three areas of social robot design – their physical shape, interactions, and objectives – are today completely devoted to bridging the gap between objects and subjects in order to ensure that only human-like actors will inhabit the social. We have thoroughly discussed the potential psychological implications of this approach and what kind of relational settings it produces. Alienation due to surreptitious otherness may lock users into a hallucinatory relationality, fulfilling (at least in some cases) the concerns shared by scholars, including Richardson, Gutiu, Danaher, and Sparrow. In the first section of this work, we proposed to transcend a theoretical approach to sociality focused on the distinction between humans and objects. We intend instead to adopt a framework where sociality is construed by actors.

This entails a radical change of paradigm in the design and evaluation of human–robot interactions (HRIs). Nowadays, interaction studies are mainly focused on one-to-one interactions between a human and a robot. The typical experimental setting requires that a human and a robot be isolated in an experimental context where they interact in a fixed way, with a precise schedule for the interaction and predetermined actions that the robot might perform. This kind of approach is justified by the importance of controlling hidden variables influencing the experiment and therefore controlling the experimental setting as much as possible. If successful, such an operation enables us to demonstrate the influence of a precise interaction pattern on the subjective reactions of humans. This usually

DOI: 10.4324/9781003459798-7

entails working with more than one condition in order to establish a control for the experimental group so we can look for differences between the two groups. A statistical analysis, usually ANOVA (Cardinal & Aitken, 2013), then returns a numerical output that is supposed to inform us about the influence that the variable we tested had on users. Typical dimensions measured in HRI include perceived likeability, sociability, trustworthiness, and agency (Heerink et al., 2010). Adding more than one variable to the experimental setting necessitates additional groups of people as controls. This approach is supposed to return solid, statistically significant scientific results. The problem with the current approach is that one-to-one interactions greatly reduce the opportunities for testing social robots' effectiveness in a systematic fashion. The current HRI, following one-to-one experimental settings, is actually assuming that we should measure the effectiveness of social robots on the basis of their interactions with a single human, rather than measuring their effectiveness in enhancing sociality in groups. The implicit assumption is that it is the robot that must be perceived as social, not the interactional ecosystem itself. Obviously, if we assume that the robot itself must be perceived as social, we will measure only dimensions related to human sociality, usually anthropocentric attributes such as "perceived agency". Seibt et al. (2020) recognise that, in a one-to-one interaction, these dimensions might impose a limitation because the interaction might be set up in a non-anthropomorphic way. Although this involves a fundamental further step, we need to deeply reshape the methodologies that evaluate HRIs in order to fully display the potential of social robotics. The reshaping might start from this consideration: in social robotics, we should enhance not the sociality of the robots but that of the hybrid interactional ecosystem. We contend that this is the appropriate standpoint for measuring the kind of modification SRs bring to human social systems – from a small group of people to the entire society.

This approach can finally enable us to tackle hybrid interactions. Three fundamental considerations guide this theoretical turn:

1 *Sociality is a property of a system.*
  We consider sociality to be a property of the hybrid interactional system, abandoning the current standpoint that considers sociality a property of an agent. This entails a deep reformulation of the evaluation methodologies, criteria, and measures that current interaction studies adopt. While these are focused on measuring the sociality of robots, we aim to measure how robots increase the sociality of hybrid systems. Moreover, we aim to understand what kinds of social roles (anthropomorphic or sociomorphic) robots might play.

2 *Anthropomorphic interactions are, and always have been, a subclass of sociomorphic interactions.*
Seibt and actor–network theory (ANT) have already brought us the necessary theoretical tools to consider objects as actors within the social. Social robotics pushes these considerations further by showing that talking objects resembling humans will produce social interactions that do not fall into the category of anthropomorphic sociality. Humans have always displayed sociomorphic interactions since the social eco-system has always been inhabited by non-anthropomorphic social mediators. Therefore, the distinction between subjects and objects, in the social world, has always been surreptitious. What becomes post-anthropocentric is neither the robot nor the human being but the space of interaction.

3 *Analysing hybrid interactional systems, we will discover new forms of social acting.*
If we broaden the view of sociality to recognise it as a property of a system, what is currently considered a failure in HRI can be under-stood in terms of another framework. Here, if a HRI is failing from an anthropomorphic point of view, it would no longer be important whether the robot is perceived as social. Of more interest would be how such interaction failures might be a driver for sociality between humans in hybrid systems, since they constitute one of the ways in which the robot can be at the centre of the social interaction. Moreover, humans might "borrow" social behaviours and cues from artificial agents, hybridising their own interactional setting – although this would not produce alienating effects on humans since the artificial agent would not mimic human behaviours, preventing the psychological implications highlighted in a previous section of this manuscript from happening. The fact that humans might display non-anthropomorphic social behaviours given the presence of artificial agents in their soci-ality is strictly related to the discussion in the section on networks of meaning. When artificial agents hybridise sociality with sociomorphic relational patterns, they will contribute to the socio-symbolic organ-isation of society. As the first talking artificial social agents, their contribution to the socio-symbolic organisation will be qualitatively different from that of any other social actor currently mediating social systems since they will contribute to the particular linguistic process that enrols, from the plasma, the signifiers that give meaning to social actions. As thoroughly discussed in Chapter 4 in relation to Maturana and Varela, autopoietic units enter into a structural coupling with the socio-symbolic system. They adapt their internal organisation to the outer social environment in order to signify social experiences (or they

increase the system's metastability). This implies that the changes to the socio-symbolic organisation of social systems due to artificial agents will have an impact on the organisation of each social actor, eventually reshaping social practices. Therefore, we claim that talking artificial agents will shape the socio-symbolic organisation in a hybrid way, deeply modifying human sociality.

The starting point for this research agenda is to shift attention from one-to-one interactions to robot–group interactions, rethinking the evaluation criteria in light of point 1 above, namely that sociality is a property of a system. This will be the first step in conceptualising evaluation and design methodologies for hybrid interactional systems in place of single interactions between human and synthetic mediators. Today, there are some approaches proposing to overcome the one-to-one experimental setting in controlled environments, although these approaches still constitute a niche in HRI. These can be grouped under the umbrella term "social robotics in the wild" (Jung & Hinds, 2018; Park et al., 2020).

These considerations lead us to put forth some possible lines of research for the analysis of hybrid interactional networks. First, we claim that widening the focus to robot–group interactions will make it possible to observe the functioning of hybrid interactional system. From point 3 it follows that an actor-robot can catalyse modifications to human–human interactions (HHIs). This point is, in our opinion, the focal difference between the notion of sociomorphing at the level of one-to-one interactions and at the level of hybrid interactional systems. On the first level, the OASIS framework developed by Seibt (2017) is a functional tool for analysing the HRI. On the second level, however, the degrees of simulation are no longer relevant since the sociomorphic setting is transferred to the network itself. Obviously, the network cannot display or simulate anthropomorphic behaviours. On that level, we might observe sociomorphic patterns of interactions between humans.

What new kinds of interactional patterns, then, might emerge from a process of negotiation that passes through a non-human actor? Although some theoretical attempts have been made to answer this question (Bisconti, 2021; Possati, 2021), the answer can only be ascertained experimentally. Meanwhile, we claim that in order to observe this possible phenomenon, we should first enquire to what extent a robot can be the spokesperson for a network. We claim that the degree of sociomorphism of a hybrid network will depend on the role that the non-human mediators have in negotiating the interactions between actors. This entails both the verbal and non-verbal semantics produced inside the network. Some experiments have been carried out, but the research in this field still lacks methodologies (Oliveira et al., 2021).

Instead of focusing on the effectiveness of a one-to-one interaction, experimental studies in systemic HRI should seek to understand:

(1) the extent to which a robot can enrol a group of actors – that is, the extent to which the robot's interaction can increase the sociality within a group of humans;
(2) the extent to which a robot can take on the role of a network's spokesperson – that is, how far it can lead the negotiation process within a network of actors.

As mentioned above, it is important to underline that interaction failures, which the literature on HRI has investigated in depth (Satake et al., 2008; Serholt et al., 2020), might be a strong driver of robot sociality if seen from a systemic point of view. One of the easiest mediation processes highlighted by ANT is when things do not go as expected, since this breaks the process of intermediation. A robot's "being a robot" is certainly one of the elements that drive the interaction between humans when confronted with a machine. The forerunners of HRI experiments (Wang et al., 2022) highlighted that Paro, even if poorly interactive, was able to catalyse interactions among elders (cuddling Paro together, talking about the robot, etc.). Therefore, from a systemic point of view, the degree of anthropomorphism is no longer what predicts how social a robot is, but how much sociality a robot can bring to a system.

If sociality is a property of a system, social robots should aim to enhance it. Systemic HRI, as a new field of studies, should enquire just what kind of systemic property "sociality" is. We can preliminarily define this property as a degree of interconnection between actors, as ANT would define it. Or, with a slight difference, we might define sociality as the volume of verbal and non-verbal communication exchanged inside a network of actors. These two definitions would capture only the quantitative side of sociality, leaving aside the qualitative. Surely, a massive street fight between gangs is a quantitatively a big social phenomenon, but as social robot designers, we would not wish to encourage this kind of sociality. Therefore, we need to understand how to measure qualitative aspects of sociality as a property of a system.

Moreover, even a successful definition of the qualitative aspects and their measures would still capture only one side of sociality, that of interactions between actors. We would be still locked in the limitations of ANT, unable to describe how social behaviours and network enrolments modify the socio-symbolic system, eventually changing the autopoietic organisation of humans. Therefore, we will also need to understand how the principles and mechanisms of socio-symbolic systems work: to operationalise the theoretical framework of socio-symbolic systemic change in measurements and design recommendations.

This pathway could eventually lead to a radical reformulation of machine design, abandoning a principle of machine social skills (so to speak) for one of machine social effectiveness in a broader sense. On this approach to design, we should not be concerned about a social robot's efficiency in displaying social behaviours. However, we do aim to make SRs effective in improving sociality. For example, European projects under the Horizon framework secured vast amount of funding for research in social robotics. The promises of this technology were impressive, and it seemed to be a solution, for example, to loneliness in elders, a growing problem in the European Union (EU). Examples of EU projects aiming at improving the social abilities of robots include Sophia, Animatas, Socrates, Spring, and many others.[1] Unfortunately, the results of recent years of research are mostly disappointing: even apart from technical deficiencies affecting, for instance, manipulation of objects or the ability to move smoothly in an environment, social robots usually fail to create long-lasting engagement with humans. Although humans may take interest in the robot for a while, this interaction rarely stands the test of time. While, in the first months, users and stakeholders are attracted by the novelty, in a relatively short time they stop interacting with the robot. We claim that this outcome illustrates our point above – that there is more explanatory and causal power at the level of the social system than at the level of individual interactions. In the first months, the simple presence of a robot in the environment improves people's social behaviours: they talk with the robot, they talk with each other about the robot, etc. When the robot is no longer a novelty – that is, when it stops *mediating* sociality because of its novelty – it becomes socially irrelevant even though its "social skills" in one-to-one interactions have not changed.

We wish to point out an issue emerging from this reasoning. The effectiveness of a social robot in the first few months (or sometimes days) is not due to its interactional abilities but mainly to what we might call "the alien effect".[2] This is something we clearly noticed during an experiment with elders in which two videos of a NAO training an elder were, respectively, administered to two groups of elders (Antonioni et al., 2021). In the first video, the NAO was interacting with an elder to help train him, and its style of interaction was quite neutral, not engaging the user with any particular verbal strategies. It simply gave instructions about physical exercises. The second video, shown to the second group of elders, showed a NAO whose interaction was very engaging and positive. For example, when an exercise was completed, the NAO congratulated the elder ("Well done! We are an amazing team") and used additional verbal strategies intended to make it likeable and socially engaging. Using a validated Likert-scale questionnaire, we measured likeability and sociability, along with other dimensions not relevant to this discussion. Through statistical analysis, we found that neither of these dimensions showed a relevant difference between the two groups (with p-value). This result was quite awkward. We also noticed that in both these dimensions, the two groups gave nearly equally (and highly) enthusiastic feedback. This led us to consider the "alien effect". Imagine that we find two aliens completely different from humans. One has six legs, while the other has seven. Administering a photo of Alien 1 to one group of participants and a photo of Alien 2 to another group, we ask them to complete a questionnaire measuring the dimension of "weirdness". We would probably derive the same results from both groups, notwithstanding the aliens' differing numbers of legs. The "alien effect", in our opinion, is deeply polluting interaction studies: without serious longitudinal studies, we cannot discriminate the alien effect from the importance of variables corresponding to "number of legs" – the variables we set out to measure.

This brings us to the same conclusion as above: if we keep thinking that sociality is a property of the robot, we cannot theoretically resolve the contradictions we find between results in lab experiments and in real-life scenarios.

So, what kind of approaches should we use to design and evaluate social robots? On the design side, we offered a theoretical turn to refocus the approach to social robots in a post-anthropocentric way. The overturn of the anthropocentric (and therefore anthropomorphic) approach might be the very solution to the psychological implications of companion robotics that we outlined in Chapter 6. The reason why the robot is alienating, uncanny, confirmative is precisely because the criteria of its effectiveness are anthropocentric. If we switch to a hybrid approach, the robot will not need anymore to be collusive to be effective. Likeability is an item that increases if the robot always agrees with the user and therefore enacts a collusive pattern. Therefore, if we consider "likeability" a criterion of the design effectiveness, we will actually design robots that will (in certain cases and only with certain users) degrade their relational abilities, as we claimed in Chapter 6. If "likeability" is not anymore a criterion to evaluate SRs' effectiveness, the designers will not be required to build collusive robots. As we can see, design and evaluation are tied in a double knot: we cannot consider one without the other. Switching from one-to-one interactions to robot–group interactions is a first step to move forward from an anthropocentric perspective. In this first step, we should consider an effectiveness criterion of how much the artificial agent is able to *quantitatively* increase the sociality of the system. On a second step, we should understand how the robot can increase *qualitatively* the sociality of the system. As said these two are pretty different things. To understand the second, we need to enquire how the semantic negotiation processes are carried out in hybrid interactional networks. In order to accomplish this objective, we cannot rely only on quantitative measurements. In-the-wild experiments in fact require that qualitative approaches, as ethnographic studies, participatory observation, focus groups, unstructured interviews (etc.) are carried out to capture all the different details of hybrid interactional networks. Rigid quantitative approaches are more meaningful when the research questions and the theoretical landscape of a field of research are clear. In fact quantitative approaches enable replication and scientific validity. On the other hand, they are much less useful to explore a new theoretical horizon. Therefore, we hope that qualitative research will assume a greater importance in the study of systemic social robotics.

Beyond social robots: synthetic social mediators

Finally, we want to point out the relevance of our discussion for so-called artificial agents, which we will more properly name "synthetic social mediators". This is the general term we use for any kind of actor that enters sociality as a quasi-other, be it embodied or virtual, for example the synthetic social mediators run by AI in the metaverse.

In fact, when we talk about social robotics, we are capturing only one distinctive instance of the current social technologies, and probably the most problematic one in terms of effectiveness. The embodiment of social robots entails difficulties related not only to interaction and social cognition but also to perception, manipulation, and movement. Social robots are the first social technologies "on the field" and are usually supposed to interact with humans in unstructured and changing environments. This entails many technical issues reducing the robot's social effectiveness. For example, a social robot supporting nurses in an elder-care facility must be able to move inside the corridor in order to reach the elder and engage in social interaction. To go from point A to point B autonomously requires the robot to perceive the environment, recognise it, predict eventual changes in it, move effectively when the scenario changes, have functional actuators for administering medications, produce context-coherent verbal language, and eventually even understand verbal language. Social robotics is the flagship technology for synthetic social mediators, but it is not currently the readiest. Such AI technologies as Generative Adversarial Networks or Google's recent Lambda are enabling effective conversational social agents such as Replika, a companion application. Or, for example, they are enabling AIs to mimic real, existing persons as deepfakes (Westerlund, 2019). Deepfakes are surely the most pressing technology, which might affect our political systems as well. These contents are generated through Generative Adversarial Networks (GANs). These are a class of neural networks capable of creating multimedia content (photos, videos, audio) that simulate real contents with extreme precision. If trained on a face, GANs are able to make it move and speak in a way hardly distinguishable from a real video (Y. Wang et al., 2022). These contents are called deepfakes.

Born mainly as pornographic contents, they have quickly moved into the political arena. Imagine a video depicting the Israeli prime minister in private conversation with a colleague, seemingly revealing a plan to carry out a series of political assassinations in Tehran. Or an audio clip of Iranian officials planning a covert operation to kill Sunni leaders in a particular province of Iraq. Or a video showing an American general in Afghanistan burning a Koran (Jahankhani et al., 2020). Similar contents were actually produced, as for example deepfakes of Barack Obama or Donald Trump. Moreover, during the war in Ukraine a deepfake of President Zelenski surrendering was released. Luckily, the quality of the video was too low to be perceived as trustworthy, but it is just a matter of time for the technology to improve and become more deceiving. GANs will likely be used to generate synthetic agents interacting with users in the metaverse. GAN-generated fashion bloggers and influencers are already a reality in China, and many are concerned that these artificial agents might be used for political reasons in a short time. EU politicians are concerned too, but the "mitigation" strategies that are envisioned are simply the limitation of such technologies, something that will be hardly doable, considering the fluidity of the digital space. The lesson learnt from social robotics will be likely applied in the metaverse, where synthetic social agents will be run by AIs without all the issues brought by embodiment, and real-world environments. The discussion carried on in this manuscript aims to apply to all the hybrid systems where there are synthetic social mediators, recognised in the form of quasi-others. The same reasoning we brought about social robotics can be applied to this wider realm. How to address the challenges brought by generative AI should move from a technology-oriented approach to a system-oriented approach. We should study how information, decisions, and mediating action can happen in a hybrid space, to understand what differences will emerge from "classic" digital spaces. What kind of networks of association, mediation processes, and spokespersons might be created in a hybrid system that is in relation with a digital environment? This issue pertains to ANT and system theory. We notice that, in this case, not only the system of relations is hybrid, but also the environment in which the actors interact is digital. This means that both sides are different from "usual" social systems and social interactions. A systemic approach might be able to precisely describe the mediation processes and offer new theoretical tools in order to ensure safety and social benefits for this new configuration of social systems.

Therefore, in hybrid interactional systems, be they inhabited by social robots or AI in the metaverse, we commit to adopt a systemic view of sociality. In the context of the metaverse, we aim to analyse how synthetic social mediators circulate signifiers in the social system, both searching for quantitative and qualitative characteristics of hybrid system with respect to non-hybrid ones. Psychometric scales, evaluating user perceptions, will not be any more linked to the evaluation of the likeability, sociability (etc.) of one synthetic actors. We will focus on how certain synthetic actors, enabling certain signifiers to circulate or to be enrolled from the plasma, might change the subject's perception of the social system itself. Highly interesting constructs, coming from the systemic work psychology, might be revitalised. For example, the construct of "perceived uncertainty" (Milliken, 1987) might help researchers in measuring "in-the-loop" perception of users in hybrid and pervasive environments.

Trustworthiness: property of systems or property of agents?

In the concluding section of this book, we expand our analytical lens, pondering the potential of transferring the core insights of this manuscript across another domain, that of AI trustworthiness. From a methodological standpoint, particularly concerning the analysis of the interplay between societies and SRs this work's paramount revelation is the conceptualisation of sociality as an attribute of a system, rather than an individual actor. This paradigmatic shift alters our evaluation criteria for the sociality of robots, and synthetic social agents at large. We posit that a similar paradigm can be applied to the notion of trustworthiness in AI systems.

The contemporary discourse is rife with debates surrounding AI trustworthiness and the design of AI systems that merit trust. The crux of the theoretical challenge lies in delineating the features that constitute a "trustworthy" system. This quandary spans both the perceptual realm, where we deal with "perceived trustworthiness" (Song & Luximon, 2020), and the "objective" realm, which pertains purely to the object's intrinsic trustworthiness. Readers who have journeyed through this manuscript might find the term "objective trustworthiness" intriguing. We employ "objective" to accentuate the distinction between an actor's internal state, which perceives its surroundings as trustworthy (regardless of its veracity), and the actual existence of this trait in the environment. The former is rooted in belief, warranting psychometric evaluation, while the latter exists externally, independent of actor perceptions.

Historically, interpretations of trustworthiness, be it perceived or objective, have been singularly fixated on the technology, typically an AI system. As highlighted by Aquilino et al. (forthcoming), perceived trustworthiness has predominantly been construed as a human actor's judgement of an AI system, drawing parallels with human trust anchored in the concepts of competence, benevolence, and integrity. Readers might discern a familiar pattern we elucidated for sociality: trustworthiness is exclusively attributed to the AI system, it emanates from an anthropocentric perspective and is gauged solely in a one-to-one interaction. The prevailing literature on trustworthiness is gradually pivoting from this traditional dyadic interaction between human and machine, where trust attributes are borrowed from human-centric trust literature, encompassing benevolence, competence, and integrity. This forms the foundational layer of trustworthiness analysis. Competence is encapsulated as the "trustee's capability to fulfill the truster's needs" (McKnight et al., 2002), aligning with the technical prowess and reliability of autonomous agents. Conversely, the remaining elements are tethered to an anthropomorphic perception of AI, as both inherently suggest intentionality. This is vividly articulated by McKnight and colleagues (2002), who characterise benevolence as the "trustee's inclination and motivation to act in the truster's favor" and integrity as the "trustee's adherence to honesty and commitment".

Recent scholarly endeavours adopting a systemic perspective herald a more holistic comprehension of human–AI teaming. This approach transcends the examination of isolated components, extending its gaze to encompass environmental ramifications. Such a paradigmatic evolution is palpable in research centred on human–autonomy teams (HATs). With the progression in AI's capabilities and modalities of interaction, these entities have transcended their erstwhile instrumental roles, emerging as collaborative counterparts. Models, such as the distributed dynamic team trust framework posited by Huang et al. (2021), endeavour to capture the multifaceted nature of HATs, assimilating both perceived and technical trust facets across a dynamic temporal spectrum.

In our perspective, we aim to surpass existing theoretical paradigms by shifting our emphasis towards the perceived trustworthiness of a hybrid interactional ecosystem, encompassing both human and non-human actors. This deviation from prior methodologies is of significant consequence: our effort is no longer confined to understanding how systemic variables, including environmental determinants, influence the binary relationship between an individual human and a technological counterpart. Rather, our endeavour is to comprehend how the inherent structure of an assemblage of human and non-human actors – a hybrid system – affects the perceived trustworthiness of all constituent actors. It is important to recognise that when entities collaborate to achieve a mutual goal, the trust placed in the collective ensemble becomes of utmost significance, surpassing the trust in discrete participants. The effectiveness of a collective is not simply an aggregate of individual contributions but is also deeply shaped by the inter-relational dynamics within its constituents.

Such insights necessitate a profound re-evaluation of our metrics and loci of trustworthiness measurement. Pioneering trustworthiness as a non-anthropomorphic construct entails conceptualising attributes that can be non-metaphorically applied to AI agents and hybrid systems.

The discourse surrounding the objective, or intrinsic, trustworthiness of AI systems has evolved significantly in recent scholarly literature. This conceptual development can be traced back to contributions such as the EU Commission High Level Expert Group's "Ethics Guidelines for Trustworthy AI" (AI HLEG, 2019). The term "trustworthiness" has been increasingly employed as an encompassing descriptor for the requisite attributes of AI, yet its precise operational definition remains a subject of academic contention. While certain foundational attributes of a trustworthy AI are well-established, as evidenced by the characteristics delineated in the aforementioned Ethics Guidelines (including robustness, fairness, societal well-being, among others), and the references embedded within the forthcoming EU AI Act, a comprehensive exploration of the EU's trajectory in AI policy-making remains outside the scope of this discussion. For a more exhaustive exploration, scholars are directed to works by de Pagter (2023) and Renda (2020). Our concern is the propensity to attribute trustworthiness solely to the AI entity, rather than considering the broader interactional system within which it operates. It is imperative to recognise that technology does not function in a vacuum; the trustworthiness of AI is a product of the dynamic interplay between the system and its external environment. The domain of AI–human teaming, therefore, emerges as a critical area of study: here trustworthiness is inherently linked to collective objectives, necessitating a departure from dyadic conceptualisations. Within this paradigm, the intrinsic "robustness" of an AI system becomes a secondary consideration; the primary focus shifts to the efficacy of interactions among different AI systems, other technological entities, and human actors in achieving collective objectives.

However, this holistic perspective appears to be underrepresented in the current standardisation processes for AI. An examination of "ISO/IEC TR 24028 Information technology – Artificial intelligence – Overview of trustworthiness in artificial intelligence" reveals a predominant emphasis on the technical attributes of AI, encompassing robustness, bias mitigation strategies, model explainability, and resilience. From a philosophical standpoint, the prioritisation of robustness as a cornerstone of trustworthiness warrants critical reflection. Defined as the unwavering performance of a system, irrespective of external variables or adverse conditions, robustness exemplifies an AI's resistance to external influences – a notion antithetical to interactional attributes. An interactional conceptualisation of trustworthiness should, by contrast, emphasise the adaptive capacities of AI in relation to its environment. As previously articulated, there exists a distinction between efficiency and effectiveness. Thus, an AI system deemed technically trustworthy may not necessarily engender trustworthy interactional systems if its design is predicated on insularity from its environment.

In summary, the focus should shift from solely creating trustworthy AI to developing trustworthy hybrid interactional systems that include various actors. This gradual shift aligns with the AI's impact within its ecosystem, aligning with the EU's directive to standardise an impact-based approach, as outlined in the AI Act. This perspective is consistent with the principles of Industry 5.0, as presented in the EU White Paper of 2021 (EU Commission, 2021). Trustworthy ecosystems require a comprehensive assessment of AI's impact on communities and societies. Therefore, designing trustworthy hybrid systems is essential for a human-centric approach to AI technology development.

This work has taken a comprehensive journey. Starting with a foundational framework on social robotics and interaction studies, we have moved to a new analytical dimension: hybrid interactional systems. We've integrated insights from various fields, including philosophy, sociology, engineering, and psychology. Social robots serve as a prime example of the changes our social systems will soon face. The integration of synthetic social mediators into our systems offers a chance to redefine the relationship between humans and their environment. We hope that readers will recognise this work as a step towards a new theoretical framework for understanding interactions between humans and non-humans in our emerging hybrid societies.

Notes

1 EU-funded projects under the Horizon framework can be found at https://cordis.europa.eu/projects/it
2 I owe this definition to Emanuele Antonioni.

References

Antonioni, E., Bisconti, P., Massa, N., Nardi, D., & Suriani, V. (2021). Questioning items' link in users' perception of a training robot for elders. In H. Li, S. S. Ge, Y. Wu, A. Wykowska, H. He, X. Liu, D. Li, & J. Perez-Osorio (Eds.), *Social robotics*. ICSR 2021. Lecture Notes in Computer Science, vol 13086 (pp. 509–518). Springer. https://doi.org/10.1007/978-3-030-90525-5_44
Bisconti, P. (2021). How robots' unintentional metacommunication affects human–robot interactions. A systemic approach. *Minds and Machines, 31*(4), 487–504. https://doi.org/10.1007/s11023-021-09584-5

Cardinal, R. N., & Aitken, M. R. F. (2013). *ANOVA for the behavioral sciences researcher*. Psychology Press.

de Pagter, J. (2023). From EU robotics and AI governance to HRI research: Implementing the ethics narrative. *International Journal of Social Robotics, 1,* 1–15.

European Commission, Directorate-General for Research and Innovation, Breque, M., De Nul, L., Petridis, A. (2021). Industry 5.0 – Towards a sustainable, human-centric and resilient European industry, publications office of the European Union, 2021, https://data.europa.eu/doi/10.2777/308407

Heerink, M., Kröse, B., Evers, V., & Wielinga, B. (2010). Assessing acceptance of assistive social agent technology by older adults: The Almere model. *International Journal of Social Robotics, 2*(4), 361–375. https://doi.org/10.1007/s12369-010-0068-5

High-Level Independent Group on Artificial Intelligence (AI HLEG). (2019). *Ethics guidelines for trustworthy AI*. European Commission, 1–39.

Huang, L., Cooke, N. J., Gutzwiller, R. S., Berman, S., Chiou, E. K., Demir, M., & Zhang, W. (2021). Distributed dynamic team trust in human, artificial intelligence, and robot teaming. In *Trust in human-robot interaction* (pp. 301–319). Academic Press.

Jahankhani, H., Kendzierskyj, S., Chelvachandran, N., & Ibarra, J. (2020). *Cyber defence in the age of AI, smart societies and augmented humanity*. Springer.

Jung, M., & Hinds, P. (2018). Robots in the wild: A time for more robust theories of human-robot interaction. In *ACM Transactions on Human-Robot Interaction (THRI)* (Vol. 7, Issue 1, pp. 1–5). ACM New York.

McKnight, D. H., Choudhury, V., & Kacmar, C. (2002). Developing and validating trust measures for e-Commerce: An integrative typology. *Information Systems Research, 13*(3), 334–359. https://doi.org/10.1287/isre.13.3.334.81

Milliken, F. J. (1987). Three types of perceived uncertainty about the environment: State, effect, and response uncertainty. *Academy of Management Review, 12,* 133–143.

Oliveira, R., Arriaga, P., & Paiva, A. (2021). Human-robot interaction in groups: Methodological and research practices. *Multimodal Technologies and Interaction, 5*(10), 59. https://doi.org/10.3390/mti5100059

Park, C. H., Ros, R., Kwak, S. S., Huang, C.-M., & Lemaignan, S. (2020). Towards real world impacts: Design, development, and deployment of social robots in the wild. In *Frontiers in robotics and AI* (Vol. 7, p. 600830). Frontiers Media SA.

Possati, L. M. (2021). *The algorithmic unconscious: How psychoanalysis helps in understanding AI*. Routledge.

Renda, A. (2020). Europe: Toward a policy framework for trustworthy AI. In D. Markus, F. P. Dubber, and S. Das (Eds.), *The Oxford handbook of ethics of AI* (pp. 650–666). Oxford University Press.

Seibt, J. (2017). Towards an ontology of simulated social interaction: varieties of the "As If" for robots and humans. In *Sociality and normativity for robots* (pp. 11–39). Springer.

Seibt, J., Vestergaard, C., & Damholdt, M. F. (2020). Sociomorphing, not anthropomorphizing: Towards a typology of experienced sociality. In M. Nørskov, J. Seibt, & O. S. Quick, *Culturally sustainable social robotics*. Frontiers in Artificial

Intelligence and Applications (Vol. 335, pp. 51–67). IOP Press. https://doi.org/10.3233/FAIA200900

Satake, S., Kanda, T., Glas, D. F., Imai, M., Ishiguro, H., & Hagita, N. (2008). How to approach humans?-Strategies for social robots to initiate interaction. *Proceedings of the 4th ACM/IEEE International Conference on Human-Robot Interaction, HRI'09*, 109–116. https://doi.org/10.1145/1514095.1514117

Serholt, S., Pareto, L., Ekström, S., & Ljungblad, S. (2020). Trouble and repair in child–robot interaction: A study of complex interactions with a robot tutee in a primary school classroom. *Frontiers in Robotics and AI, 7,* 46.

Song, Y., & Luximon, Y. (2020). Trust in AI agent: A systematic review of facial anthropomorphic trustworthiness for social robot design. *Sensors, 20*(18), 5087.

Wang, X., Shen, J., & Chen, Q. (2022). How PARO can help older people in elderly care facilities: A systematic review of RCT. *International Journal of Nursing Knowledge, 33*(1), 29–39.

Westerlund, M. (2019). The emergence of deepfake technology: A review. *Technology Innovation Management Review, 9*(11), 39–52.

Index